Life
Story

60 years of love, books, dogs, and gardening

Jonathan Gregory

Dedication

For Douglas Adams, who gave us the answer to the meaning of life, the universe and everything which, as everyone knows, is 42.

This is a work of fiction.

Any resemblance to any people, living or dead, is totally accidental and unintentional. Apart from the real people I have shoehorned, Zelig-like, into this story.

Books by the same author

The Gemini and Flowers Mysteries
Country Life
Family Life
In Real Life
After Life
Unsporting Life
Wild Life
Private Life
High Life
Street Life
Time of Your Life

*

The Tall Timber Trilogy
Tall Timber Falls
Tall Timber Rocks
Tall Timber Lives

*

Other Books
Nice People
Accidental Murderer
The Stone Age
Just a little ghost story
Another little ghost story
A little Swedish murder
The King's House
Who dun it?
The Wood Man
Cow Boy
Life Story

1958

Chapter One

Fletcher's Cross, 2nd April

When Mrs Roscoe died, there was much speculation in the village of Fletcher's Cross as to who would buy Wisteria House, her beautiful home overlooking the green. The late Mrs Roscoe had been a widow, and her three sons had all died in the last war; two pilots and one army officer, all buried in distant lands far from the Cotswolds. With no one in the family left to take over her estate, it was put on the market by her executor and purchased rather quickly by a man called David Manners.

After some repairs, additions and redecoration, a large pantechnicon from Pickford's Removals arrived on the second of April, and three men in brown overalls spent five hours unloading and unpacking its contents, including a baby grand piano. A van arrived at the same time from "Stroud Drapery Limited", and two other men, wearing white overalls, carried in several packets wrapped in brown paper and started hanging curtains at all the windows.

There was a black Rover 90 parked in the driveway that led down the side of the house to the garage in the converted stables, but no one had yet seen Mr Manners. Several neighbours strolled by at different times of the day, hoping to catch a glimpse of the elusive new owner, but he remained out of sight.

At three in the afternoon, the blue Pickford's lorry drove away and, an hour later, the men hanging the curtains also left. With the front door closed, Wisteria House once again appeared unoccupied and silent.

It was a large, detached property with four bedrooms. Built of honey coloured Cotswold limestone, it had white window frames and a pale blue front door and fronted the street overlooking the village green.

There was only a long, narrow border filled with mature lavender and rosemary bushes separating the building from the pavement. A vast wisteria grew all over the frontage, and the buds indicated this would be another bumper year for blooms later in the spring. There were smaller cottages, both thatched and tiled, on each side, with long strip gardens behind them. Wisteria House had a wide lawn between two overgrown herbaceous borders at the rear, that led to a large ornamental pond and two weeping willow trees. Between these stood a magnificent magnolia *Grandiflora*, in full bloom. Behind each border was a beech hedge. The garden then spread out to the left and the right from the pond, behind the neighbouring gardens. On one side was a walled vegetable garden, another vegetable patch, and an orchard. On the other was a heather rock garden leading, via a rhododendron walk, to an area of beech woodland of about six acres, with mature trees. That spring, the ground under these was carpeted with bluebells.

The back door was on the side of the house and led to the kitchen through a scullery. A gravel path ran along the rear of the property, going around a stone terrace in front of the French doors in the living room. Behind the garage was a potting shed and an area of rough grass where washing could be hung out to dry, a boxed area for compost, a coal store, dustbins and an old, rusty incinerator for burning garden, and other, rubbish. Everything outside was neglected and in desperate need of attention.

Rumours, spread by Sam Jones, the plumber who had worked on the house for the new owner, said that it now had two bathrooms, with the new one attached to the master bedroom, and the whole kitchen had been "ripped out and filled with new stuff, the likes of which I ain't seen before. All cream coloured, with a zinc draining board so long you could lay a man out on it, you could."

But sadly, for the regulars at the Fish and Trumpet, Sam Jones hadn't met the owner, nor could he say if there was a Mrs Manners. All his negotiations had been done through the estate agency handling the sale for the executor.

On the third of April, on the public notice board outside the post office, passers-by saw that a hand-written sign offering work at Wisteria House for a cleaner three days, and two half days, a week, and a gardener full time, had been pinned up with four brass drawing pins. The house had a substantial twelve acres of garden in total; all in need of an expert's hand.

Mrs Roscoe had run out of money before she ran out of life, and the huge garden had not been attended to for five years.

Several working women dropped in letters that same day offering their services: one, Lottie Nolan, a war widow, aged thirty-nine, with a twelve-year-old daughter, was invited, via a postcard slipped through her door later that afternoon, to the house the next day. She left an hour later, having secured the position as cleaner, starting the following Monday. As Lottie kept herself to herself and never visited the pub, any information from that source about the new owner would have to wait.

On the fifth of April, the vicar, The Reverend James Fallow, popped into Wisteria House to welcome the newest member of his flock at eleven in the morning; well, the visit was to establish whether Mr. Manners would be joining his flock. He left an hour later, with a cheque for twenty-five pounds towards the church restoration fund, and a smile on his face, but as he didn't gossip and also never visited the pub, that route for information was also denied the good citizens of Fletcher's Cross.

At three that same day, young William Forman, a twenty-three-year-old gardener and handyman, knocked on the back door of the house overlooking the green, and he left two hours later having been offered the gardener's job. He and Mr. Manners had strolled around the grounds and discussed all that needed to be done, and Will could understand why the new owner wanted a full-time gardener.

Will went to the pub that night and told everyone he had been given the job. He informed his rapt audience that Mr. Manners was a widower, aged about thirty, a writer, and he owned a piano and a dog called Flush.

'What books do he write then, Will? Ain't heard of an author called Manners,' asked Bert Hoskins, the local butcher.

'He has a pen name, that's what he said. Connor Lord.'

Hoskins gasped, 'He writes them spy books? Bloody hell, them's a bit racy.'

Will, basking in the attention, replied, 'That's what he said. He had a box full of the latest one, *The Bern Assassination*, on his kitchen table. Oh, and that kitchen. All new, it is. Really beautiful.'

A couple of the farmer workers sniggered at this comment. Men were not really meant to notice things like nice kitchens, but Will was made of stronger stuff and turned to one and said,

'Michael Strong. If your good woman 'ad one of them kitchens, she might even lift her skirt for you more than once a season, you miserable bugger.'

'Now then, now then, lad, none of that sort of talk in 'ere,' said Barney Young, the publican. The pub regulars laughed. They all knew that Mrs Strong seldom let her husband, a tight-fisted, dull man, partake of the pleasures of her flesh; and, for most of them, they thought it would take a lot more than a new kitchen to change that.

The conversations started up again around Will as his fellow villagers chewed over this new arrival. A writer. Of spy novels. That was new and very exciting.

The first Sunday after Manners' arrival in Fletcher's Cross, people wondered if he would finally show himself and come to church. Just before matins started, a tall, good-looking dark-haired man strolled up to the church door, tied a black Labrador to the timber bench outside and took his place in the back pew. He was dressed in a well-cut grey suit, with a white shirt and blue tie. He looked very masculine and fit, a bit like a young army officer. There was a ripple of voices through the congregation, who were slightly aggrieved at him sitting at the back, as Mrs Roscoe had always sat in the front pew. They would have to turn their heads, which was very unseemly, to catch a glimpse of the new owner of Wisteria House.

St. Stephen's was a small Norman church, with wonderfully painted, if slightly faded, frescoes of crusading knights on its creamy-white walls. That Sunday, there were red tulips arranged in a vase on a tall stand next to the lectern, and the fine Georgian silver cross on the altar stood gleaming in the sunlight, with large oak candle sticks on either side. The church held about two hundred souls when a service was fully attended, as it was that day. The vicar progressed up the aisle, the service started, hymns were sung, and a short sermon about neighbourliness delivered. By the time most people had come out of the church and shaken the vicar's hand, Mr. Manners had collected his patient dog and was walking home. He had thanked the vicar for a fine service, put five shillings in the collection dish, and left before the rest of the congregation had had time to see him properly.

People walked by the front of his house on their way home, and heard piano music being played, but there was no sight of the man himself again that day. The members of the congregation that had sat in front of him duly reported he had a fine singing voice, a welcome addition to the performance of hymns at St Stephen's. From then on, people passing Wisteria House would often hear him playing the piano. He played classical music and some jazz; Manners found it helped him when he was stuck on a particular chapter or scene in one of his stories, but his neighbours didn't know that.

At eight o clock on Monday morning, Lottie Nolan arrived for her first day of work. Will Forman also arrived at the same time, and Manners sat with them in the kitchen having a cup of tea and discussing what needed to be done first. Then he left Lottie to get on with her work, as he retreated to his study to write, and Will fetched the scythe to start cutting away five years of neglect in the rear garden. At eleven, they all met up again the kitchen for another cup of tea and a biscuit, and Lottie left at four having cleaned the bathroom, dusted the living room, made the double bed, hoovered the whole of the downstairs apart from the study, and cleaned up the kitchen.

At five, Will knocked on the back door to tell Mr. Manners he was done for the day. He had cut away all the long grass on the lawn, burnt it in the incinerator, and weeded the gravel path and the long border at the front. The author appeared from his study and said, 'See you in the morning, Will. You did a good job today. Damn good, in fact. Thank you.'

'Thank you, sir.'

'Now, I must take Flush for a walk.'

'Have a nice evening, sir.'

Will walked home, smiling. He had not been this happy, ever, in his whole life. He had always worked since leaving school at fifteen, but his previous employers had never praised his efforts on their behalf, nor paid him as well as Manners did. He was earning fifteen pounds a week, a full two pounds over the national average, and more than most of the farm workers in the neighbourhood. But it wasn't the thought of the money that made him smile; he liked the words of appreciation even more.

By Wednesday evening, the lad had cleared the weeds and dead stalks from the two long herbaceous borders that ran down each side of the lawn. He showed Manners all the different perennials that were coming up, and which would provide a splendid show in the summer.

'Them's peonies, and we've got lupins, phlox and those are asters. Hollyhocks coming up at the back. Got some red-hot pokers too, them's next to the cat mint. And echinops; them's the ones with those deep purple blue spiky flowers, bit like Christmas tree decorations. Lovely, they are. Bees love 'em. And those are verbenas. Give a good splash of colour in the summer.'

David was very impressed by Will's knowledge. On Thursday afternoon, as he was leaving for the day, Manners said, 'I think I'd better show my face in the pub tonight. Get to know people. Good idea, do you think?'

'Ay, well, there's lots who want to know more about you, sir.'

'Will, please call me David. I'm far too young to be called "Sir" all the time. Okay then, I'll pop down.'

Will hesitated then said, 'Yes, David. In front of others too?'

'Not a problem for me.'

'Oh. Right then. Well, might see you in the Fish and Trumpet then…'

*

'Evening, Sir. What will it be then?'

'Oh, a pint of Stoat's Revenge, I think. I've heard good things about it. From young Will over there.'

'Ah, you must be the mysterious Mr Manners then.'

'Hardly mysterious, just new. And it's David, please.'

'Then I'm Barney, Barney Young. I'm the landlord here.'

'Nice to meet you, Barney.'

The publican drew the pint into a glass tankard with a thick side handle and placed it in front of his new customer. Flush had arrived with Manners and was now sitting at his feet by the bar, ignoring a Jack Russell bitch that belonged to an ancient farm worker called Grumbley, which was intent on flirting with him. The snug bar, the only bar in The Fish and Trumpet, was dark, had low ceilings, walls yellowed with nicotine on which hung a few horse brasses and hunting prints, a couple of bench seats, and small wooden stools grouped around a few round tables. There were about twenty men in drinking that night, and no women. David tasted the beer, nodded his approval, slid a half-crown over the bar and said, 'Join me, please, Barney.'

'That's kind of you, David. Just a half for me. So, have you settled in?'

'I think so. It's a fine house and I'm being well taken care of.'

'Nice to hear. We didn't see much of Mrs Roscoe after her hubby died, but they used to pop in on Sundays after church sometimes; well, Mr Roscoe did.'

'I heard they lost three sons in the war. Desperately sad.'

'Did you serve?'

'I did my national service after it ended. Two years. You were in the army?'

'Yes. Burma.'

'Christ, that was a tough one.'

Barney nodded. 'Lost a lot of friends out there.'

'Brutal indeed. Do you think this one was the war to end all wars?'

'Doubt it. Men have a habit of fighting each other, don't they? Even in peacetime.'

'Sadly true. Did you lose a lot of chaps from around here?'

'Several, as most weren't viewed as essential workers despite the farming.

We fed a lot of folk from the land around here, and got land girls in to help out. Not many conscientious objectors here either. But most of the younger lads, like you, have done their national service. Will did. That's where he got shot.'

The boy had a very slight limp, but Manners had been far too polite to mention or ask about it.

'Really? I noticed his gait but thought it might be a recent accident.'

'Nope. One of his fellow conscripts dropped his rifle, cocked, and it went off and shot Will in the calf. Got an honourable discharge… he were only in eight weeks.'

'Oh, poor chap. Well, some of the men I was in with should never have been allowed near any form of weaponry.'

Barney nodded knowingly. He had come across a few English soldiers in the war who had been more dangerous than the Germans or the Japanese.

'Is it true you write them spy books?'

'About John Wright, secret agent? Yes, those are mine.'

'Very popular they are, even if they are a bit… a bit…'

'Racy?'

'That's the word.'

Manners laughed.

'They are a bit. But I think we need a bit of fun after all the rough times we've had. People need to relax and escape. I mean, go to London and the damage from the last war is still everywhere. We might have stopped having our food rationed, but it will take many, many years to rebuild our cities. For most people, books are an escape from that. I'm just doing my bit to help them.'

'And making a bob or two in the process?'

'Indeed,' chuckled his new customer.

'Will said you've got a new one out. I'll get it out of the library when it comes in. They are always a bit slow with new books.'

Manners smiled and said, 'I can let you have a copy. As the author, I get a few free examples. I'll drop one in if you'd like it.'

Barney nodded happily. He loved books but he loved free things more.

He was famous in the village for being a bit tight-fisted.

'Oh, that would be very generous of you, David.'

'Not at all. This is a very good pint, by the way.'

'You should try the Widow's Ale too. Same brewery, but a bit more on the stout side.'

Will joined them at the bar, ordering another half of that very beer.

'I'll pay for that, Barney, and another one of these for me. Will's done an excellent job so far. Deserves a little bonus.'

'Well, that's nice of you, sir,' said his gardener. He just couldn't bring himself to call Manners David in public. It wasn't right. But he added a 'Thank you.'

'Not at all. And you were right. This is a very good pub.'

The three men nodded to each other in mutual agreement and started to discuss the village. All around them was the murmur of conversation from the farm workers and the other regulars, all of whom had kept one ear open to listen to the new arrival. Now they settled down to do some serious drinking, assured he wasn't that strange after all. He drank pints and had a nice dog, which was basically all it took to be accepted in the village. He wasn't unduly posh, or snobbish, wasn't loud, and treated his new employee with consideration. The same could not be said of some of the richer people who lived around Fletcher's Cross.

It was a small hamlet of about one hundred and twenty homes, with a duck pond and a war memorial to the fallen of the two world wars on the green, and a baker, a butcher and a greengrocer. Will's mother worked behind the counter in the bakery.

The post office was next to the pub, and a small garage and petrol station could be found on the road out towards Stroud. A red telephone box had been placed next to the greengrocer, as most people still didn't have a telephone in their homes. As the village didn't have a policeman either, there was a blue police box behind the war memorial, for everyone to use in case of emergencies. Due to low farm wages, and the general level of countryside poverty, many of the cottages still had outside lavatories, called privies, and their owners took baths in free-standing zinc tubs in front of their open fires in the living room.

To the north lay the Cotswolds Hills, with three dairy farms, and to the south a couple of arable farms and a pig farm. The eastern side of the village, across the green and the main road from Manner's house, bordered the Langham estate, owned by Lord Langham, who bred horses and daughters; the poor man had eighteen of the former and five of latter, all aged between twenty and twenty-seven. Lady Langham, an exhausted woman of fifty-three, was seldom seen, preferring to paint insipid watercolours of the landscape around their stately home than take part in the social scene of the county.

There had been a couple of other large houses built on the outskirts of the village during the thirties, both originally occupied by retired civil servants, back in England after long stints serving His Majesty in colonial positions. These two houses had been sold to local social climbers; one to Morgan Hudd, who made his money from coach tours, and the other to Peter Franklin, a builder of very poor-quality houses, who had done well, and was still doing so, out of the post-war construction boom. Neither of these families were very popular with the locals, and the staff turnover was high, due to their general rudeness and because they paid low wages.

Hudd and Franklin competed with each other constantly. By buying bigger and more expensive cars; younger and more expensive wives and their current passion, racehorses. Both had bought thoroughbreds which had run at Cheltenham and other events, but neither had yet had a winner. They used different trainers, each of whom in turn was getting rich as the nouveau riche pair fought for supremacy.

There were two other villages nearby. Studley Combe, to the south, had about two hundred houses and cottages and was where the local doctor and vet had their practices.

Combe Weston, between Fletcher's Cross and Stroud, had another hundred or so homes, and a riding school.

All three had a handful of council houses for the low paid and those farm workers who didn't get a tied cottage that went with their job, and there was a bus service that passed through them six times a day, connecting them to Stroud, with its grammar school, big public library and railway station to all points north, east and south.

This the was the village and its surroundings that David Manners had chosen for his new home. Only time would tell if he had made the right decision.

Chapter Two

On the day following his visit to the pub, Manners walked to the post office at three in the afternoon, carrying a parcel wrapped in brown paper. It contained the first three handwritten chapters of his sixth book, *The Paris Affair*. The parcel, Mrs Hobbs the postmistress noted, was addressed to a Miss Bottomely in Swindon. She was a free-lance typist, wheelchair-bound following a childhood bout of polio, who worked on all David's manuscripts. He would try and write at least two or three chapters a week in his neat handwriting; then make changes, and the annotated documents were sent to Miss Bottomely, who returned them a week later, beautifully typed, double spaced, ready for further changes, along with the original manuscript, so he could compare the two, if necessary.

Each book ran to roughly thirty chapters, so took about ten to twelve weeks to write one. He only started writing when he had worked out the plot in detail and done copious research to make sure his locations, guns, and scenarios were accurate. The whole thing, with numerous changes, was returned to Swindon for the final version to be typed up, which was then delivered to the publishers in London five weeks after that.

David had kept all his original manuscripts, right from the start. Once a book was finished, he would tie all the pages up with string, and stuff it into a drawer in his desk, along with his notes and research. As the number of books rose, he started storing these papers in old banana boxes. At Wisteria House, which had an extensive attic, he stored those there so they wouldn't get in the way.

Flush sat outside the post office as David handed over the parcel, then the two of them strolled back to Wisteria House via the butcher's shop, where David bought two large lamb chops, a pound of pork sausages, six eggs and a marrow bone for his dog.

Bert Hoskins wished him a good day, then turned to Alf, his assistant, after his new customer had left and said, 'Funny, isn't it? With all that money, you'd have thought he'd employ a cook.'

'Men do cook sometimes, Mr Hoskins.'

'Not rich blokes, they don't.'

This was also a matter of some discussion with several of the women of the village. They had waited to see if another advertisement for a cook would appear on the notice board, but it hadn't. Lottie Nolan, taking care of the flowers in church one Saturday afternoon, had been waylaid by the vicar's wife, who asked about the cooking arrangements. Lottie, a bit nonplussed at being tackled over her employer, had simply replied, 'He does it all himself. Leaves the dishes, pots and pans for me to wash up, but he does all his own cooking, Mrs Fallow.'

'How very odd. What sort of things does he cook?'

'Bacon and eggs, or sausages, for breakfast most days. He has a sandwich for lunch and a proper meal in the evenings. He likes chops, stews, shepherd's pie, that sort of thing. Loads of vegetables too. Prepares them all himself. I've offered but he seems to enjoy doing it. Says it helps him think and write.'

'Plotting his books whilst peeling his spuds?'

'Yes, Mrs Fallow.'

'How very, very odd.'

David Manners' habit of cooking for himself became a regular topic of conversation in the greengrocer too, and it was during one of these gossip sessions that Lady Langham's cook overheard the discussion and duly reported it back to the staff at Langham House. Lady Langham's personal maid, Michelle, mentioned it to Her Ladyship, who informed her husband and daughters over dinner that night that, not only was there a highly successful and single male writer living in Fletcher's Cross, but he also spent much of his time cooking for himself. This led to a lively discussion about men cooking and the suggestion that perhaps an invitation to dine at Langham House should be issued.

The following day Lady Langham wrote to David, inviting him to dinner that Saturday. A brief note arrived the next day accepting this invitation. Lord Langham, up in town for a debate in the House of Lords, dropped into Foyles bookshop before returning home, and purchased the first of John Wright's adventures, *The London Connection*, to make sure he knew a little about their prospective guest's writing. By the time the train pulled into Stroud railway station, His Lordship was slightly nervous; the book was indeed, "quite racy, m'dear" as he informed his wife that same evening.

At twenty past seven on Saturday evening, a black Rover 90 swept up to the front of Langham House, and the author got out. He looked very handsome in his dinner jacket and black tie and strolled slowly over to the edge of the gravel to study the herd of fallow deer that had free range over much of the parkland. He then walked to the front door and was welcomed by the butler, before being escorted into the drawing room where the family and other guests had assembled. All five of the Langham girls looked very pleased that their mother has invited him, before it dawned on each of them that they were now in competition for the first good-looking, healthy, wealthy young male the area had seen for quite some time.

Manners greeted Langham, who insisted he call him Richard, shook hands with Her Ladyship, "my wife, Susan", then chuckled and said, 'Johnny, good to see you.'

He shook hands with John Ridge, RN retired, who had been a godsend when it came to information for one of his books.

'Oh, do you know each other?' asked Lady Langham.

Ridge nodded and said, 'David contacted me at the Ministry two years ago, asking lots of questions about naval activities and terminology for "The Portsmouth Attack". We've kept in touch ever since. Love the new one, by the way. Thank you.'

David had sent him a free copy as a gift for past assistance.

'Glad you like it. How are you enjoying retirement?'

'Bored out of my mind, old boy. Started play golf, of all things. You've never met my wife, have you? Loraine, this is David Manners.'

'I love your books too. But why write under another name?'

David laughed. 'Well, they are a bit rude in places and, when I wrote the first one, I had assumed I might end up with a career teaching English, and few schools would approve of one of their staff members writing racy fiction. However, it was so successful, I've been able to make a living just by writing, so it doesn't really matter. Also, Connor Lord does sound a bit more exciting than David Manners.'

He was introduced to the five girls and the other two guests, Major Jack Hartley and his wife, Mary, old racing friends of the Langhams. After a bit of inconsequential chat covering the weather, the upcoming Cheltenham races and the problems of free-roaming deer and dog walkers, the butler announced that dinner was served and they went into the dining room. David had been placed between Florence Langham, the eldest daughter, and Chloe, the youngest.

As they enjoyed the fine oxtail soup, Her Ladyship asked, 'David, how does it work with book sales nowadays? Do most people buy them or borrow them?'

'Ah, interesting question, Susan. When it comes to hardbacks, it is mainly the libraries that buy them so people can borrow them. People tend to buy the paperbacks now, but not the hardbacks. Usually. But some people still only buy hardbacks. They look better on a bookcase.'

'I did,' said Richard Langham. 'Jolly racy too. I bought the first one.'

'Will you buy the second?'

'Oh yes, I think so.'

'That's wonderful. The good thing about writing books men enjoy is that they tend to collect them. Once they have found a writer they like, they keep coming back for more. I'll let you have the new one. I have some copies at home.'

His host, who was always acutely aware of money due to his excessive daughters and the costs of running his stately home, replied, 'Oh, that would be very kind.'

Florence asked, 'How old were you when you wrote the first one, Mr. Manners? You must have been jolly young.'

'David, please. I wrote the first when I was eighteen and finished it during my first year at Oxford. Sent it off at Easter to an agent, and he sold it to Norman Hilton, of Hilton Press. I was very lucky. They were not looking for new writers at the time, and there was still a shortage of paper after the war, but they did like the book. Then one of their other authors was having problems finishing his next novel and they had a gap in their publishing schedule, so agreed to take mine after all. *The London Connection* came out in October that year, as I was just starting my second year. I had just turned nineteen by then. And I'd written most of the second book during the summer holidays.'

'Good grief,' muttered Richard Langham.

David laughed, 'Yes, I was a bit precocious back then.'

The soup dishes were removed and the next course served, a small Welsh rarebit. Chloe asked, 'Mr. Manners. Why aren't you married?' John Ridge sucked in his breath. He knew the whole story – well, he knew the author had been married. David blushed, dabbed his lips with the white napkin and stammered, 'Um, Oh, gosh. I was. Once. I am a widower.'

'Oh Lord, I am sorry,' responded Chloe. 'I've put my foot in it again.'

'Not at all. Why should you have known? My wife died in a road accident eighteen months ago. We were living in Oxford at the time, and her parents lived in York. She drove up to see them for Christmas. I was going to go up later by train. It was during that period of terrible fog at the end of '56. She was hit by a lorry on the Leeds to York Road. It was very quick.'

There was a stunned silence then Susan Langham said, 'You poor man. How awful.'

He smiled sadly. 'She was carrying our first child. Yes, it was hard. It's one of the reasons I moved to Fletcher's Cross. I found Oxford too full of memories of her to carry on living there. And we had passed through here on our honeymoon. I remembered seeing the house covered in wisteria at the time, so when it came on the market, it seemed like fate to me. It's a lovely place.'

'We never knew the last owners,' said Susan.

'Well, if you ever want to look around, feel free to drop in. The garden is still a terrible mess, but I've found a chap to take care of it, even if it might take all summer to knock it into shape again.'

The difficult topic moved over, there followed much discussion about gardens, growing vegetables (The Langhams had grown many during the war) and the difficulty of finding good staff. Petunia, the next eldest daughter, giggled and said in an exaggerated, shocked tone, 'Is it true you cook, Mr Manners?'

Now he roared with laughter. They were just finishing off some excellent roast beef and he replied, 'Yes, I do. I enjoy it. I can make bread, bake pies, roast meats, stew things and can put together a damn good trifle, if I say so myself. Helps me sort out my plots as I work on preparing a meal. Do you cook, Miss Langham?'

'God no. Can't boil an egg.'

'Sad. You should try. Most young ladies do nowadays.'

'My dear Mr. Manners, I am not most young ladies. None of us are. We're the Langham girls.'

Her father sighed and said, 'Pet, don't be tiresome. The fact that you have been spoilt from birth does not bode well for a future with few staff. All of you girls might have to pick up a mop or a rolling pin at some point in the future.'

'I do hope not, papa,' said Gillie, the fourth daughter.

After coffee and a game of bridge, David thanked his host and hostess and drove home. He let Flush out and sat on the edge of the terrace, smoking his final cigarette of the day, as his dog ran around, sniffing the grass. They had a hedgehog visit the garden from time to time and Flush could tell it had been through while he had been trapped in the kitchen. He peed on the border then came over and David rubbed his head. 'You're a very good dog, you know that?'

He got licked heavily, Flush's way of agreeing with him. David looked up at the night sky, thinking about the dinner party. He was in no hurry to repeat the evening; he had found the girls boring and typical of their class. Slightly stupid, but sure of their own desirability. None of them had made any impression on him and, in fact, his sexual desires lay elsewhere.

For David had a secret, one that was illegal and would, if known, have destroyed both his professional career and his standing in society. For David Manners, despite his brief but happy marriage, preferred the company of men.

Chapter Three

He had discovered the true objects of his desires during his school days. Like many public schoolboys, the lack of girls and the surge of hormones made many young lads turn to each other for relief and pleasure, snatched in rude moments in the locker rooms or woodland around the establishments. Although most grew out of such acts after leaving school, David continued, finding fellow travellers during his three years at Oxford University where he read English at Corpus Christi College. Liaisons between students, whilst frowned upon and still illegal, were part of Oxford life, and he enjoyed his time there greatly.

It was during this period that he met Eve Watson, who would later become his wife. Eve was in love with an American student, one Sally Wickers, a year above her, and the three became friends. Eve knew he was a homosexual, and he was having fun with a short, very fit member of the Oxford rowing team, at the time; the boy was Catholic and would go and confess his sins every Friday. They were not in love, but very much in lust.

When Sally returned to the States, Eve was very unhappy, and turned to David for comfort. He had just had his second novel published and was subject to much attention as the reviews had been universally excellent. Another intelligent spy novel, highly entertaining, it had been well

received and sold extremely well, making David that rare beast, an author who could live on his writing. It was during a riverside walk that Eve admitted that, even though she was a lesbian, she wanted a child but could not see herself ever marring a "normal" man. David, wanting male company but knowing how dangerous such acts would be to his newfound success, suggested a compromise. He could think of being a father. They should marry and have a child, but with no expectations of regular sex between them once the child was on the way. A marriage of convenience indeed.

Such an arrangement would give both of them the respectability society demanded. Eve graduated and started working as a research fellow, and they moved into a small house together near Jericho, as friends at first.

The plan worked, and their friendship deepened. He sent off his third book to be published, which occurred during his national service and, as soon as he had been discharged, they married and, five months after their wedding, Eve announced she was with child. Once she was pregnant, they both found they were really looking forward to the event, but Eve's death put an end to this. David was very sad, but not totally bereft, discovering that being a young widower made him an even more sympathetic figure to the wider public, and might provide perfect cover for his homosexual tendencies. But he missed Eve dreadfully as a friend and mourned his unborn child. He found he could not stay in Oxford, hence his move to Fletcher's Cross.

He kept the house in Jericho and rented it out to students through his old college. Since Eve's death, he had not indulged his sexual passions; there had been several high-profile prosecutions recently and the times were not friendly at all to anyone caught performing homosexual acts. He focused on his writing and tried to ignore his more carnal lusts. However, his decision to employ Will had partly been due to the fact the boy was stunning. He had a short, stocky but very fit body, thick blond hair, blue eyes and full, very kissable, lips; his shy way of talking and his appreciation of any praise made him all the more desirable. But David would never have made a move on the lad, knowing such action would result in his career ending and him being shunned by society, if not imprisoned, had he been rejected. Instead, he transferred his lust onto the pages of his new book, as John Wright fell madly in love with a Russian temptress called Ivana Rusenski. The sixth book would prove very racy indeed.

*

At the end of April, just three weeks after David moved in, the wisteria covering the front of his new home burst into flower. It was magnificent, with hundreds of tiny pale lilac-blue blossoms, grouped like small chandeliers, cascading from the thick branches of the climber. The scent was stunning, and people would stop and admire the blooms, inhaling their fragrance. The flowers hung down over the windows and the lintel of the front door. They covered the whole of the frontage, from the eaves to the lavender beds below.

The rest of the village also came to life. Late flowering daffodils nodded across the green and in big clumps by the war memorial. Other wisteria, and some early-flowering clematis, burst into bloom, and David was very happy he had chosen this village. It was delightful, peaceful, and, for the most part, he was left alone to write.

Occasional visitors were either ignored if Lottie wasn't there or turned away if she was. He found her a warm and very kind woman, and they were soon on first name terms. She would tell him about her daughter, who had a lovely singing voice, and sometimes her late husband, who had been killed in the desert, fighting Rommel. She missed him very much and like most war widows, couldn't see herself remarrying; able bodied men in their forties were rare birds in those parts. She had also lost a brother in the war, on the beaches at Normandy.

She supplemented her income (also fifteen pounds a week) from working for David by knitting, and her employer ordered two pairs of thick socks; he tended to just wear socks when he was writing as they were more comfortable. In summer he went bare foot.

The new book was going well, and he wrote from eight-thirty in the morning until four-thirty every afternoon, with a short break at noon to walk Flush and eat a sandwich. When he put down his pen at the end of a day's work, he would call for his dog and the two of them would go and find Will to tell him to go home, his day done. The lad had been making excellent progress. After the initial cutting back of the overgrown grass at the rear, he had allowed it to green up again, then started on it with the hand mower. It was beginning to look like a pristine lawn should, but it would take the full growing season to bring it back to any semblance of its former glory. David could image the last family playing croquet on it, or simply lounging about in lawn chairs, having drinks before dinner. After the grass and the flower borders, Will had turned his attention to the horribly overgrown vegetable garden, which was surrounded by a high stone wall. He had stripped away the detritus of several years, discovering a fine patch of rhubarb, a fig tree, a wonderful rosemary bush and a vast patch of mint.

He had cut back the things that needed cutting, double dug the soil, asked for, received and mixed in a trailer load of fine horse manure, and had started planting for that year's crop. Cabbages, along with leeks, carrots, early lettuce, parsley, sage and thyme. Runner beans were under glass, and radish seeds had been set in long lines in one of the beds. He and David had discussed the new owner's preferences for food and, as a result, they had purchased two peach trees, already trained to grow against a wall, and two pear trees, which Will espaliered professionally. David was very impressed by his knowledge and enthusiasm for the garden.

One afternoon, when he had finished writing for the day, he went out into the garden to tell Will to stop working. It was slightly later than usual, as he had been finishing a rather dramatic scene in his story. The lad was digging behind the walled vegetable garden, in what used to be a large potato patch. He was working on the main area but had weeded two long strips along one edge and David could see dozens of tiny green shoots sticking out of the ground.

'What are those… oh, are they asparagus?'

'Yep. I've just weeded them to give them a chance. Not sure how they'll do this year. They should have had a bit of manure last autumn.'

'Oh, I love asparagus. That's wonderful. Let me know how they are coming along. Um, I came to tell you to finish for the day. Your mother will be waiting for you with your tea.'

'What time is it then?'

'Five fifteen. Don't you have a watch?'

'Got me granddad's old Timex but forgot it today. I'll just finish this bit then head off home. Thanks for telling me. Do you really like that asparagus stuff? Don't it make your wee smell funny?'

David roared with laughter. 'It does indeed. Horrid. But they are good to eat.'

'Not worth it if you ask me.'

He finished the potato patch the next day. The next phase involved sorting out the area around the huge ornamental pond, like the rest of the garden, currently very overgrown. Will started by carefully dragging out much of the weed, freeing trapped fish and amphibians as he did so. He cut back masses of yellow irises and bullrushes, trimmed back the branches of the tall willows overhanging the water, and cut the grass around the edge, which revealed large flat stones, interspersed with bugle, ferns, and other damp loving plants.

Buried inside a clump of irises that had been at the centre of the pond, Will discovered the head of a fountain, a simple flat bronze affair with a dozen holes in it. The two of them spent hours trying to find the switch to turn it on, and finally located it in the potting shed, under the light switch. It was a black oval, which one clicked right to turn on, and left to turn off. Will switched it on and they went back to see the water gently sprinkling out.

'That will be good to get the water circulating, David. It will oxygenate the water for them fish.'

'Excellent. Surprised it still works. Couldn't have been used for years.'

David purchased some fine water lily plants, to protect the now exposed goldfish from any passing heron, and a wrought iron and timber garden bench so he could sit by the pond and watch the aquatic life. He would rest there and have a cigarette sometimes when working on a plot problem, with Flush standing quietly by the water's edge, tail wagging slowly, as he followed the fish as they moved around beneath the surface.

He had jumped into the water the first week after a frog, so now his master said, 'No, Flush, you can't go in, old boy.' The dog woofed at him and resumed his study of the goldfish.

David didn't smoke much but enjoyed each cigarette when he did. Will sometimes joined him if he had a question about the ongoing work, and he often cadged a fag off his employer; his mother didn't approve of smoking so he could not do that at home, he said.

They would sit there and chat, which Will found difficult to begin with, being shy by nature. David slowly extracted the information that he had started gardening because of his favourite teacher in the village school, who had raised roses. She had asked him to help her out in her garden and encouraged him to learn more about plants and how to grow and care for them. He had made a living doing this, as well as supplementing his income with a few handyman tasks; mainly a spot of painting – front doors and windows, or creosoting fences.

'Course, they didn't pays as much as you do,' he said one afternoon.

David had laughed. 'I have always believed in a fair day's pay for a fair day's work, and you work damn hard, Will.'

'Ta. I like this garden. Very much.'

Another time they were sitting smoking when Will frowned and said, 'Here, I've been meanin' to ask you. Why do you have that big, lumbering Rover 90? A young guy like you should have an MG or something else sporty.'

David had nodded then smiled sadly. 'I did. I used to have one. Eve, my wife, was driving it when she died.'

Will looked shocked and was horribly embarrassed he had stumbled onto his employer's painful tragedy. 'Oh. Oh. Oh sorry.'

'You weren't to know. It's alright. It was a lovely car too but offered no protection against lorries. I'd love to have one again but I just can't bring myself to buy one. I got the Rover as it is solid, built like a tank. Chances are, if I hit something or something hits me in that, I'll survive. Also, Flush finds the back seat very comfortable, and the MG's canvas roof used to leak on him in wet weather.'

Will still looked very contrite about bringing up such a sad moment in his life, so David offered him another cigarette to show there were no hard feelings and turned the subject to television instead. 'What do you like watching?'

Will looked down at the grass and said, 'We don't have one. A television. Not yet. I want one but mum's against it.'

'Oh. That's a bit boring of her. Well, if there's anything special you ever want to watch, let me know. You are more than welcome to come over and watch mine.'

While many houses now had television, they numbered less than sixty per cent of the village. The coronation two years earlier had been the trigger for most families to consider getting one of the heavy sets with their tiny black-and-white screens, but they were still expensive and most people rented theirs. The tell-tale aerials attached to chimneys and eaves clearly indicated those who had a television and those who didn't. David had one in a walnut cabinet with a roll down door to keep it hidden when he wasn't watching it.

'Ta, that would be great...,' said his gardener.

It was during one of these pond-side chats that Will brought up the subject of chickens. Like most gardens, theirs was plagued by snails and other insects, especially the vegetable garden.

'Could we get some chickens? I could let them out in the morning and put them to bed before I go at night, and they would keep the garden free from a lot of pests. Or at least, cut down the number of snails.'

They were watching a thrush bang a snail to bits on one of the stone slabs by the pond. He added, 'Those birds do their best, but there are so many.'

'Do chickens eat snails?'

'They do. Well, them buggers eat anything really. We could put a coop up just outside the vegetable garden wall, at the rear, so they don't wake you up.'

David called around the farms and found a farmer who was willing to sell him six Little Sussex hens and a cockerel (as Will had informed him those particular brown speckled hens were great layers) and they duly arrived, along with a brand-new hen house. After that, they had a steady supply of fresh eggs, and a very quick reduction in the number of salad plants lost to snails. David was settling into his new life and enjoying it greatly.

A pair of swallows arrived and set up home under the eaves of the garage, and a wren nested in the wisteria by the front door. He liked birds so noticed these arrivals with real pleasure. He got into a regular routine; he would work, walk Flush, play the piano and go to the pub a couple of evenings each week and began to get to know the local characters. They were mainly farm workers, manual labourers, or they worked in the shops in the village. The local doctor popped in from time to time, a sad looking man called Threlfall. He would sit with a small brandy, reading the Times, then go off home to his wife and small children in Studley Combe. Apparently, he found the noise of their playing very disturbing. David fitted in well with this crowd and had given Barney his free copy of the fifth Wright adventure which had been passed around the other customers. They all agreed it was very "racy" indeed. Other regulars included the local car mechanic, an ex-RAF man. He had serviced Spitfires in the war and spoke as if he had won the Battle of Britain single handily. There was a retired cobbler who still mended the farm workers' boots from his shed in the back garden, and a man who bred Pembroke corgis, one of which he had sold to the Queen Mother. He often brought his favourite to the pub, and Flush would roll over and play with the bitch under the table.

David became a regular at church on Sundays and now didn't rush off after matins as he had done that first Sunday. He would stand, dog at his heels, chatting to the other members of the congregation, or the Reverend Fallow himself.

Will was always in attendance with his extremely pious mother, a tight-lipped woman who constantly looked as if she was in pain. Lottie Nolan also went, with her pretty daughter, destined to break hearts across the county. Her name was Flora; she was twelve going on thirty, and she flirted with the village boys mercilessly during the services. On one Sunday, the regular organist was ill, and Fallow asked if David knew how to play the organ.

He had played the instrument at his school, so he covered for the sick man and managed to get through the service without any problems, although the congregation was a little surprised when he played them out to *All I have to do is dream* by the Everly Brothers, a big hit on the radio at the time. Afterwards, Fallow said to his verger, 'It could have been worse. He could have played *Great Balls of Fire* by that dreadful Jerry Lee Lewis chappie.'

David would now greet people warmly on his walks with Flush to buy his provisions in the local shops. A lot of people had dogs, so his Labrador made new friends on a daily basis. And people liked David as he wasn't loud, miserable or arrogant, so he was accepted as a new and rather pleasant addition to the community. He had written to Lady Langham after the dinner and thanked her, included a copy of his new book for her husband and, at the beginning of May, reciprocated, inviting the couple, along with the Ridges and two old friends from Oxford to a dinner party, his first at Wisteria House. For this event, he recruited a couple who ran a small catering business in Stroud to cook and serve, so he could play the perfect host. It was a glorious early summer evening, so they had drinks on the stone terrace before eating in the dining room. Although the cook had provided the first and main courses, he made the trifle, which everyone agreed was excellent.

The whole evening went off without a hitch, although Lady Langham spent in inordinate amount of time recommending her daughters to her host. He told her that sadly his publishers' timetable meant he had to focus on finishing his book before he could indulge in more social activities. David's new life seemed to being developing as planned, when an event took place that threatened everything.

Chapter Four

The news spread quickly, and Lottie Nolan was the one who
brought it to Wisteria House early in the morning of the twenty-
ninth of May. A man had been found murdered in the next village,
possibly a victim of a burglary that had gone wrong.
There had been a few break-ins that month, both in their village
and in Combe Weston, where the dead man had lived. The local
policeman, based in Studley Combe, had cycled over and visited
David the week before and told him to be on the lookout for
strangers, and recommended window locks and bolts for the back
and front doors. Only a bit of money and silver had been taken
before, when the homes had been unoccupied; this was the first
death.
The next day, the villagers of Fletcher's Cross were shocked when
two police cars arrived outside Wisteria House and, a little bit later,
Will had been escorted from the back garden in handcuffs. The
rumour mill went into action and one of the more lurid tales
suggested he had been having an illicit relationship of an unnatural
nature with the deceased. David had been walking Flush when Will
was arrested, but Lottie rushed out to inform him of the shocking
event as soon as she saw him returning. He drove to Stroud
following the arrest, calling his solicitor first to go in and help the
poor boy. The police inspector interrogated Will for two hours
with the lawyer sitting next to him, then came out of the interview
room and asked to speak to Manners.
'The lad says he was with you at the time of the murder.'
'When was that?'
'We think about twelve noon on the twenty-eighth of May. The
body wasn't discovered until the late afternoon by his neighbour,
but somewhere between eleven and three on the twenty-eighth.'
'Then Will is telling the truth. He was in the kitchen of my home
eating a ham sandwich with me and Lottie Nolan, who is my

cleaning woman, then went back to clearing the ground under the apple trees. He was there until five.'

The inspector nodded then frowned.

'Now then, Mr. Manners, it has been suggested that William Forman is a homosexual. Which, as you know, is against the law. What is your relationship with him? You can see why I'm asking.'

David sighed and said, 'Of course, Inspector. If I was his lover, I would be happy to give him an alibi.'

Inspector Jones went bright red. He had not expected such a response. He was a short man, and David towered over him.

Manners went on, 'However, I am a widower, not a homosexual. And neither is Will Forman. He is a good-hearted, honest, hard-working lad, a churchgoer who looks after his mother and helps many people in our community. Also, he has, as they used to say, stepped out with several of the local girls or so I am told. I would not have employed him had I thought for one moment he was a shirt-lifter.'

'I see…' stammered Jones. 'Well then…'

'Who suggested he was?'

'This other lad. Nat Hunter. Our only witness. He says he saw William Forman leaving the property just after one that afternoon.'

'Then I suggest you talk to him again. Because it is impossible that Will was anywhere near Combe Weston at the time. Now, can I take the lad back to Fletcher's Cross. His mother must be beside herself with worry. I can assure you he won't run away.'

'Um, yes…'

David reached for a pad of paper on the desk and wrote something.

'This is my telephone number. I would be most grateful if you could let me know what is happening so I can tell Forman. To put his mind at rest.'

Jones took the note and nodded, rather intimidated. After thanking the solicitor for his help and asking him to send his account, David drove Will home again. The poor lad was very shaken by his experience, as he had never been in any trouble with the police before.

'Why did they think it were me?'

'Because someone said they saw you but it was at a time when such a thing would have been impossible.'

'Who?'

'A man called Nat Hunter. Do you know him?'

Will nodded sadly. 'We were at school together. He was a bully and tormented me almost daily. Then he got nicked and went to prison for two years for stealing. So, the police believed him over me?'

'If he's been in trouble before, he probably knew what to say to get them to look at you in the first place. What a bastard.'

'Thank you for helping me.'

'Not a problem, Will. I only hope they find Hunter again and he hasn't done a bunk.'

'I'm not a… well, you know what they accused me of…'

'I know that too.'

'I feel so ashamed.'

'You are entirely innocent. When they catch up with Hunter again, your name will be cleared.'

When they got back to the village, David drove Will home, but when they got there, there was a small leather suitcase, and a pair of boots outside on the doorstep. When Will tried to open the front door, it was locked.

He looked at David in total confusion and stammered, 'I … I don't have a key. It's never locked.'

A window opened on the first floor and his mother stuck her head out and said, 'You're not coming back in here, my lad. It's disgusting, it is.'

'Mum! I'm innocent.'

'That's not what they're saying in the village. On your way. You are no longer my son,' she shouted and slammed the window shut. Will just stood there, mouth open, in a state of shock. 'What? What do I do now? Mum!'

But the cottage window remained shut. David picked up the suitcase and said, 'Right then, grab your boots. You're coming home with me. There's a spare bedroom or two at my place.'

'Oh, sir, that ain't right…'

'I'm not having you being homeless, Will. Come on.'

Lottie was still working when they got back to the house, and David explained what had happened. She made them a cup of tea, tutting around saying how unfair Will's mother was being. The lad sat at the kitchen table and cried his eyes out, so Lottie put her arms around him and gave him a hug. David wished he could do the same.

After tea, he led the boy up to one of the back bedrooms and helped him unpack, before Will fell on the bed and slept, exhausted by his ordeal.

At six thirty, Inspector Jones telephoned to say that they had arrested Hunter on the platform at Stroud railway station, trying to get away to London. He had admitted killing the man in Combe Weston and making up the story about Will. He gave David a little more information than was normally available to the public then said, 'Tell the lad I'm sorry. He just sounded so convincing, did Hunter. We should have checked his record a bit better before taking Forman in.'

'You might have ruined his life, Inspector, so yes, maybe next time you should.'

The next day Will worked in the garden, keeping away from the front of the house and the village in general. But that evening David suggested they go to the pub and face everyone.

'We must put an end to any gossip, Will, else you'll never be able to live here happily again.'

Will was very nervous about doing this, but his employer insisted. When they arrived at the Fish and Trumpet, Barney Young was pulling a pint but stopped when he saw Will and shouted out, 'Now then, now then, we don't want his sort in here, thank you very much.'

'Now what sort is that, Mr Young?' replied David.

'He's a bloody arse bandit.'

'Ah, I see.' David turned to the pub's other drinkers. It was pretty full that evening, with people coming for the latest gossip.

'Is that what you all think then?'

There was a murmur of agreement, and a few "ayes". David stood by the bar and said, 'Right then. Let me put you all straight. Last night a man was arrested in Stroud for the murder in Combe Weston. His name is Nat Hunter. Ah, I see a few of you know him. Maybe you too went to school with him. A nasty piece of work. According to Inspector Jones, he has admitted being the burglar of your homes, the killer of Reginald Munrow and being a male prostitute as well. A more disgusting individual you could not hope to find. And he tried to put the blame on poor William here, whom you all know to be a decent, hard-working, clean living young man. Now explain to me why you, who have all known Will his whole life, would take the word of a murderer and a thief over his?'

There was an embarrassed silence, then an overweight farmer by the name of Barry Snood muttered, 'Well, there's no smoke without fire.'

David snorted with derision. Snood had supplied the horse manure for his garden, so he had heard him talking shit several times before, but not like this. He looked at the fat man and replied, 'Really, Mr. Snood? Well, I can assure you there is absolutely no smoke and no fire at all in this case. You should all be ashamed of yourselves. He is your friend and neighbour.'

'But he wouldn't date my daughter, Rosie,' muttered Snood.

David turned to Will and asked, 'Which one is Rosie?'

'That fat girl who works at the greengrocer.'

'The one with an arse like a hippopotamus and a moustache?'

The pub roared with laughter.

David went on, 'Christ, not even Grumbley would ask her out.'

The old farm worker shouted out, 'Oi might if she shaved…'

More laughter. The tension had been broken. As Snood shouted out, 'Oi, you can't talk about my daughter like that…' David turned to Barney Young and said, 'Are you going to serve us, Barney, or do you believe there is no smoke without fire too?'

'No. Sorry, Will, stupid of me. Of all of us. A pint of Widow's? On the house. For both of you?'

He really must have been feeling sorry, as he never offered free drinks.

'Thank you, Mr Young.'

Two of Will's friends came over and clapped him on the back and led him, with his pint, to their table. David lent on the bar and took his beer from the landlord, who said, 'That were a good speech. Put us all in our place, that did.'

'Bloody right. That poor lad's been put through the wringer. You know his own mother has thrown him out?'

'I'd heard that. Well, in the circumstances… but hell, it sounded so disgusting.'

'If it had been true, it would have been. Hunter was visiting Munrow on a regular basis for sex for money and killed him when he wouldn't give him more after the last time, according to the inspector. Then he remembered Will from his school days and shopped him to the police when someone said they had seen Hunter near the victim's house.'

Barney sighed. 'You hear of these things in the papers, but you don't think of them going on around here, not in the Cotswolds.'

David and Will left the pub at ten thirty. Will had had several pints bought for him by friends trying to make up for the damage and he was very drunk. He was violently sick on the village green and again in the cloakroom lavatory back at the house.

David finally managed to undress him and get him into bed just after eleven. He placed a metal bucket by the boy's head on the floor by the bed, in case he was ill again. Just as he was leaving the spare room, Will said, still very drunk, 'Thank you for believing in me, sir. But it was true. I am a shirt lifter. And I love you, sir,' before he fell asleep.

David went downstairs and let the dog out. As Flush trotted around the lawn, David lit a cigarette and inhaled deeply. So. Out of the mouths of babes and drunkards. Will would probably not remember saying that in the morning, but he would. The question was, what would he do with the information?

Chapter Five

The full story was on the front papers of all the newspapers the next day, Saturday. There was even a picture of Hunter, who did look slightly like Will, but the killer had a cold, hard look in his eyes. The report made it clear that no other people were involved and, as far as the police were concerned, they had got their murderer.

David woke at seven that Saturday morning, went downstairs and put the kettle on the Aga before letting Flush out. He made tea, then fed his happy dog, fresh from checking the rear of the house was safe from squirrels, hedgehogs, and other monsters. David went and let the chickens out then made some toast, and sat alone at the table, reading the Times, which also covered the murder, but in a less dramatic way than the Mirror. He got both papers delivered every day. One for the real news, the other to get the "man-in-the-street's" view of events, useful when writing the more working-class characters in his novels.

Through his parents, both dead, he had had a very upper-middle class upbringing, with his late father being a doctor in Salisbury. His only real contact with boys from a different background had been in the army.

Oxford, even in 1949, when he gone up, had been all ex-public-school men and women, with only a very few from grammar schools in attendance. He needed to understand how other classes thought to make sure his stories were reasonably accurate; hence he had the popular paper delivered daily.

When he had submitted his first manuscript to his agent, he had written an explanation as to the sort of person he thought would enjoy it. This was, in itself, a novelty at the time. Authors seldom considered the sales targets of their books. But David had realised early on that most book buyers were women, who enjoyed romantic fiction or Agatha Christie-type murder mysteries.

Men, if they bought books, tended to buy classic Sherlock Holmes or Ryder Haggard novels, old style adventure stories. He wanted to write modern spy novels to encourage men to try something new. As his publishing company had been struggling with just this problem, they decided to go ahead with his book for that reason as well. His success led to some jealousy at his college, both from the dons who would have killed for such sales, and his fellow students, most of whom dreamt of writing a best-selling novel, but were still waiting to put pen to paper. Nothing is more irritating to a dreamer than someone else making their dream come true, especially through their own efforts.

The main reason the book had been a success, apart from the wild, "racy" seduction that had covered two of the middle chapters, had been the ending.

His secret agent, Captain John Wright, RN, had tracked down the villain, a war racketeer called Samson, to a remote farmhouse in Wales. Samson had run guns to the Irish, girls to brothels in London and Birmingham, and made a fortune through rationing in England and Germany, often at the expense of the government and ordinary citizens.

Once captured and in handcuffs, Samson had sneered at Wright and said, 'My dear Captain, if you think for one moment that I will ever see the inside of a prison or even a courtroom, you are very much mistaken. One call to my solicitor and he will contact a huge network of people who owe me favours. You have no conception of how corrupt this country is, or how many people are in my debt.'

Wright had replied, 'And you misunderstand my mission, Samson. I am not here to take you in. I am here as your judge, jury and executioner', and had shot him in the head.

Such a breathtakingly brutal cold-blooded killing by one of Her Majesty's civil servants had been shocking, even if the readers had privately cheered his action. The ending had been viewed as very controversial, but it had made the book famous, or infamous, and sales had gone through the roof, in England and especially in America. War profiteers seemed immune from prosecution, so having one shot dead, even if unarmed and in handcuffs, was poetic justice to many.

David had done a lot of research to make sure the book felt genuine, including talking to many ex-servicemen.

Their resentment towards the spivs and war profiteers ran very deep. Many had come back after the war to find their homes destroyed, their jobs gone and their families split up. The struggle to start again had been very tough, and to see people who had never served lording it over everyone else caused much resentment. That was why David had made his first real villain such a man.

He had started on the follow-up before the first was published and completed the next one after he had finished his final exams at the end of his third year, in 1952. It was the same year another British author, Ian Fleming, also published a spy novel, *Casino Royale*, about a certain James Bond. There were now two famous secret agents on the bookshelves.

David achieved a double first, and an invitation to visit an old don in his comfortable rooms near the Ashmolean library. There the elderly man, a recruiter for MI6, had tried to persuade Manners to join them, but he refused.

His leanings, he knew, could prove problematic, so he turned down the charming offer and went to console Eve, still mourning her lover, who had just departed across the pond to New York.

David was brought back to reality by the arrival in the kitchen of a very green-looking Will, still feeling a bit rough after his night of drinking. He had had a wash, and smelt of Wright's Coal tar soap, David's favourite. It had inspired the name of his hero. The boy sat down and asked, 'Did you put me to bed?'

'I did.'

'I don't remember a thing. Was I ill?'

'Several times.'

'Oh. I am sorry.' He looked horribly embarrassed.

'Don't worry, we've all drunk too much from time to time. Are you hungry?'

'I am. Funny that, ain't it? I'm starving. Have you got any bacon?'

David got out the big frying pan and, ten minutes later, they were both eating bacon, eggs and fried bread washed down with a fresh pot of very strong tea. Will, munching happily through the feast, asked, 'What do you think I should do about my mum?'

'Wait for her to come around and ask for your forgiveness.'

'She won't do that. She's too proud.'

'Well, you can't go back and beg to be let in again. Leave it a day or two. She'll hear through the grapevine that you are totally innocent and she will have to apologise.'

'What if she doesn't?'

'You stay here.'

'That's not right. I work for you.'

'It's not a problem for me. No one else is using that room.'

The boy finished his breakfast and was silent for a while then said, 'Thank you for helping me. Getting that lawyer and everything. I don't know what would have happened if you hadn't been here. I'll pay you back.'

'Don't be daft. You might work for me, but you're one of the few people I know around here. That I like anyway. Couldn't abandon you to the wolves, could I?'

'Most would have. Look at that lot in the pub. They thought it was all true.'

'I thought you couldn't remember last night.'

'I remember that.'

David smiled. 'Nothing else?'

Now Will looked at him, startled. 'Why, what happened? What did I do?'

'You did nothing.'

'What then?'

'It's what you said.'

The boy sucked in his breath, looking very worried. He asked, very softly, 'What did I say?'

'That you loved me.'

Will closed his eyes, wishing the floor would open up and swallow him down to the depths of Hell. He went bright red and stammered, 'I…I don't…'

'It's all right, Will. I like you too.'

Will looked at him, not understanding what he was hearing. Then David smiled and winked, and the boy, realising his world was not coming to an end, looked confused.

'But you were married…'

'My wife liked women. I like men. It was a marriage of mutual convenience.'

'Oh.'

'Do you have any experience?'

The boy shook his head, very hot, red and deeply confused about having a conversation out loud about such private lusts and thoughts. David reached across the table and gently took his hand. 'Do you want some?'

Will looked up at his employer through thick eyelashes and bit his bottom lip, then nodded.

'Then I'd better lock the doors and take you to bed…'

*

Two hours later, Will lay in the crook of David's arm, hugging his chest. They were both saturated with sweat, as Manners had closed the bedroom window in case anyone outside had heard their exertions. He had one arm down Will's back and his other hand was stroking the lad's hair. 'I didn't hurt you, did I'

'It were a bit painful the first time, but nice though. Better the second time.'

David chuckled. 'It was. I hope you want some more.'

'What, now?'

'No. Tonight, or tomorrow, or the next day.'

'So, you want to keep… doing it with me?'

'Yes, of course I do. Why? Did you think I just wanted to have you then ignore you?'

'No, but… you're a posh bloke. I'm just a working lad.'

'You're a wonderful man and I want you very much indeed.'

Will lifted his head and kissed David's chin. 'Good.'

David ran his finger down the boy's back, then down the back of his left leg. He felt the indentation from the bullet wound that had given Will his slight limp. Pulling back, he looked down and at the other side of the calf.

'It went in and out. Right through,' said his new lover.

'It doesn't look bad. Bet it hurt like hell.'

'It did. Left me with a limp. That leg's a bit shorter now… the muscle is
anyways.'

'Have you tried lifts in your boot?'

'What are they?'

'Extra thick soles, or a thicker heel.'

'Um, no. Only got two pair of ordinary old boots. I'm not rich.'

'I know. Well, it's something to think about.'

He reached for his cigarettes and lighter, then balanced the Bakelite ashtray on his chest. He lit one, inhaled, then held it close to Will's lips so he could smoke too. The lad blushed; this was another, very intimate, first for him, sharing a cigarette in bed with a partner.

They smoked the cigarette together, then David crushed it out, moved the ashtray back onto the bedside table and touched the boy again. Will sighed happily, as David was now stroking him between his legs.

He said, 'Um, I don't know much about you.'

'Like what? You can ask me anything.'

'Well, how old are you? I'm not good at judging that.'

'I'm twenty-seven. Always looked a little older than my age. And you are twenty-three. You look younger.'

'That's cos I were a virgin, weren't I.'

'Well, that ship's sailed.'

They both laughed. It had indeed, very happily, gone on its maiden voyage, never to return.

Will grinned and kissed him again then said, 'Yes. Oh, something else. Apart from your wife, you never mention your family. Why is that?'

'I don't have one. I did once, of course. But I was an only child, like you. My father was a doctor in Salisbury, and mother did charity work, like most middle-class housewives. They were killed in the Barnes rail crash in 1955. They had gone up to stay with old friends in Windsor, gone to a show in town then, as they were returning to Windsor late that night, their passenger train was hit by a freight train. Thirteen people died, including my folks.'

'Oh God, first them, then your wife. You've had a rough time.'

'It was difficult, yes. I miss them. They were good people, but they would not have approved of this. Me falling for a man.'

'You're fallin' for me then?'

'You know I am.'

'Good. You have a beautiful body. I like it.'

'Good. I like yours too.'

He stroked the boy's bum, and Will giggled, 'This is nice.'

'It is, but we can't lie here all day. I think we need to get up and have a bath. Poor old Flush needs a proper walk.'

Will looked shocked. All this sudden intimacy was a little overwhelming.

'I can't have a bath with you.'

'No? Then I'll pop in first and you go in afterwards.'

They met up again downstairs. David handed Will a brown envelope and said, 'With all the excitement, I forgot to give you your wages.'

'But I missed a day.'

'I'm not going to dock you a day's pay because of that damn policeman.'

'Well, I'll go and do a bit of weeding in the vegetable garden then. To make up for lost time. I can't come for a walk with you anyway. That would seem odd.'

'It would. For now. And we'll have to be careful while you are here. Make sure it looks as if both beds have been slept in else Lottie will wonder what we are up to.'

Will went red again. 'You want me to sleep with you at night?'

'Would you like to?'

The boy nodded, then grinned shyly. David kissed him on the lips and said, 'Good. That's settled then. I'll be back in about an hour. Come on, Flush, sorry you've had to wait.'

The black Labrador bounced around the kitchen happily, holding its leash in its mouth. David extracted it, then opened the back door and they went off, down the drive and along the road towards the church. He felt exhilarated and very happy but was also aware they would have to be very discreet if this liaison was going to work.

He walked along the lane to the churchyard, then followed the wall to the gate, which stood open. There were bunches of early summer flowers tied to each of the posts. The Reverend Fallow came out from the church and walked down the path to greet him. He had a wedding that day at two thirty, and he had just been checking everything was in order before having a light lunch at the vicarage.

'Good morning, Mr. Manners.'

'David, please, Vicar. Are you working today?'

'A wedding this afternoon, that's all. Do we see you tomorrow?'

'Of course. I enjoy your sermons.'

'Ha. Well, it's nice of you to say that. Not sure the rest of my flock would agree with you. I'm sorry to hear you've had some problems this week. How is William doing?'

'He was very shaken up by the whole event, and his mother has thrown him out, which seems damn unfair if you ask me.'

'Oh Lord, has she? Where is the poor lad staying then?'

'I've put him up at Wisteria House. He's there most days working anyway, so it seemed the simplest solution to the problem. Actually, it might work out well, as I often have to go up to town and can't take poor Flush with me, so he can keep an eye on him and it is safer to have the house occupied in case we have more burglars in the district.'

'I do hope we don't. That was very generous of you, … David. I'll have a word with his mother when she comes to church tomorrow and see if I can get her to change her mind. Knowing Betty Forman, that won't be easy.' He looked at his watch and added, 'Well, I must get on. Enjoy your walk. Goodbye, Flush.'

The dog had been happily nuzzling the vicar's hand, tail wagging. David did so, taking the bridleway that led to one of the farms and skirting along the bottom of a field of dairy cows. Flush ignored them. As David walked along the track, he smiled. That had been quick thinking, but, in fact, taking care of Flush during David's visits to the big city was the perfect excuse to have a live-in gardener, even if Mrs Forman decided to let her son return home. She didn't. The next day, she sat alone at the front of the church, and refused to let Will sit next to her. He stood there looking lost before Lottie tugged on the hem of his jacket to let him know he could sit with Flora and her instead. He left church feeling very sad, then went to the pub on his own afterwards, nursing a pint for two hours before returning to Wisteria House and the loving arms of his employer.

His fellow regulars at the Fish and Trumpet thought his mother was being very unreasonable, but they all knew Betty Forman as a stubborn woman, and sympathised with her disappointed son.

Chapter Six

When Lottie arrived for work the following day, David explained the new arrangements and offered her an extra two pounds a week to cover any additional work having the boy live at the house. They had already worked out that Will would have to use the second bathroom and ruffle up his bedclothes every morning to make it look as if he was sleeping in the spare bedroom. Lottie agreed, said she thought Betty Forman was a sanctimonious old baggage, and carried on cleaning up the kitchen after their breakfast.

That Wednesday David had to go to London to meet both his publisher and agent, and he had been offered a ticket to see *West Side Story* at Her Majesty's Theatre, which meant he would be staying overnight. His excuse for having his gardener stay was thus justified, and Will took Flush to the pub with him that night to demonstrate his usefulness to the author.

The main topic of conversation in London was an offer from an American producer for the film rights to the Wright books. His idea was to set the stories in America, not post-war Europe, and have an America actor play the British hero. David rejected the whole package and suggested his agent, Murrey Holt, contact one of the British television companies instead.

Norman Hilton, his publisher agreed; he didn't want the series ruined, as David was now his best-selling author.

Manners had said, 'I'd much prefer a careful series in three or four episodes that truly reflected the books, not some sort of flash American version that wouldn't make much sense and which would lose the very essence of the stories.'

Holt agreed and said he would try his best. David spent the rest of the afternoon buying some new clothes, for him and for Will.

The poor lad's wardrobe was slim to say the least. He bought him a couple of white collar-less shirts, some white T-shirts and a pair of Lee Cooper jeans. These, produced by a British company, were England's answer to Levi's, and were just starting to take off. Will already had a pair of denim dungarees, part of any workman's unofficial uniform, but jeans based on trousers were still a bit new for the British public, outside the artistic community. David bought a pair for himself as well.

In the evening he went to see *West Side Story* with Murrey and his wife, Lillian. He had seen it when it opened the previous December, but Murrey had had a spare ticket. The show was a present for his wife's birthday, and she was more than happy for David to join them when her unmarried sister, who lived with them, cancelled due to an unfortunate attack of shingles. He had enjoyed the musical the first time and did so again.

Before leaving for home the next day, David visited John Lobb, the bespoke bootmakers on James Street, to collect a pair of brogues he had had made. While there, he spoke to the Master Last maker who had measured his feet and made the wooden forms for his shoes. He presented him with a small problem. Unwrapping the brown paper covering the parcel he'd brought with him, he showed the expert Will's oldest pair of work boots.

The man, Percy Small, sucked in his breath and said, 'Good Lord, does somebody actually wear these awful things, sir?'

'They do, but I want you to make him a new pair. Without your usual process, I'm afraid. He can't come in for a fitting. Could you manage it from the indentations in the old leather?'

Small lifted the first boot gingerly, as if picking up a nasty piece of litter.

'It might be possible. What is the story, sir?'

'He's an old friend I met during National Service. He's fallen on hard times and I want to present him with a new pair. Also, he was injured during his time with the army and his left leg is slightly shorter. Could you manage a pair where the left boot is a quarter inch higher than the right?'

'Oh, that's not a problem. We have many clients who have problems with their gait. Hmmm, well, we can do this, but they will not be up to our usual standard.'

Small was studying the boots with a very professional eye, despite their condition. He added, 'You can see the imprint of his toes and heel in this very old leather.'

'Perhaps you could make a model using plaster of Paris.'

Small gasped. 'Mr. Manners. We are not the police!'

David grinned and said, 'Could you make two pairs? One brown, one black. While you are at it?'

Small sighed, then looked at Manners, and nodded. He was an old client, like his father, and his father before him. 'Four weeks.'

'Can you send them to my home address, and I'll make sure he gets them?'

'Certainly, sir. I have an apprentice. This will be a good task for him to practice his skills.'

'Excellent. Now, let me settle my account while I'm here.'

That done, David found a taxi and headed for Paddington. When he arrived home, Will was alone. He opened the bags with his new clothes slowly, not sure how to respond to such generosity.

'These are too much, Davey. I'm not your kept man.'

Will had started calling David Davey. Neither of them was quite sure why, but David liked it. He hadn't started calling the lad Willie, for a variety of reasons.

'You need a few things. You don't have to wear them all today. Or tomorrow. But you need to have a few more clothes. I'm not keeping you; you work hard, you know you do.'

Will had looked over the new clothes, which were all much nicer than anything he had and said, 'Okay then. Not sure about the jeans, though. They are a bit…'

'Modern?'

'I'm not a hipster, you know.'

David hugged him and said, 'Oh, I know that all too well. Now, how about some fish and chips? They will be here soon, and I forgot to buy anything to cook for dinner.'

They joined the queue by the war memorial for fresh fish and chips from the van that came through the village every Thursday evening. Cooked over a coal-fired stove, they were served in newspaper. The other villagers in the queue thought it highly amusing that the owner of Wisteria House was buying fish and chips from the lorry. Apparently, the wealthier people around Fletcher's Cross just didn't do that sort of thing. They took their cod and chips back home and ate them off the paper at the kitchen table, liberally dosed with vinegar and salt, washed down by a strong cup of tea.

Will, licking his fingers to get the last of the grease and salt off them said, 'Them were lovely. Mum never allowed me to have these. Doesn't approve of food from a lorry.'

'Your mother seems to have a very strange view of this world.'

'Yeah, she does.'

'Nothing sinful about fish and chips.'

'She could find sin in a tin of Spam. Oi, Davey, I know I wanted to ask you something. Have you seen my boots? The oldest pair. Can't find them anywhere.'

David feigned ignorance and replied, 'No. How odd. Maybe Lottie has thrown them away. They were a bit... ragged.'

'I asked her but she said she ain't seen them. They were old, yeah, but they'd do for another few months in the garden.'

'You can wear a pair of my wellingtons. Use a pair of thick socks to stop them slipping off. My feet are bigger than yours.'

'Not just your feet...'

David rolled his eyes as Will added, 'Okay then, but it is strange about the other pair...'

David noticed that Lottie was looking a bit miserable on Friday morning and asked her what the problem was, concerned she might be having second thoughts about the new arrangements. It turned out that the issue was at Flora's school.

'Some of the dads go in and talk about their work, and Flora ain't got a dad to do that. She's feeling a bit down about it.'

'Perhaps I could go on her behalf and talk about my books.'

Lottie stared at him in amazement and said, 'You'd do that? For me?'

'And for Flora. It's not her fault she hasn't got a dad; it's Rommel's.'

Wednesday the following week he went to the village school and talked for half an hour to a rapt audience of children, telling them about spies and guns and wicked men and strong, hearty English heroes. It was a huge success, and Flora's standing rose dramatically.

Lottie, talking about it later with some women at the church, said, 'He didn't need to do that, did he? But he's such a kind man. Flora's ever so happy now. It's made all the difference in the world.'

David's standing in the village rose another couple of notches, and the cleaning was even more enthusiastic from then on.

*

They were laying in bed together that Saturday night, relaxing after some very gentle love making. David had brought a candle up and lit it, to create a romantic atmosphere, and he was on his side, looking down at his lover as he lay on his back, smiling happily. Will was getting more and more comfortable with his tall, dark partner, who was just running his finger over his chest between kisses. He sighed, thought of something, hesitated then asked, 'Have you been with many men, Davey?'

'One or two. A couple of boys at school and a couple of undergraduates at Oxford. But nothing like this. There was no emotion. Just lust.'

'I'd done nothing before. Before you.'

'You said. Did you want to?'

'Well, I thought about it but hadn't met anyone. All the lads in the pub, my mates, they aren't like us.'

'No, I guessed that.'

He kissed him again and smiled. 'Hear… did you give me the job because…?'

'Because I wanted you? No. You got that because you knew your stuff. I had prepared a lot of questions about gardening if you remember… I had to look them up, because I don't know a bloody thing about plants, and you answered everything perfectly.'

'Oh.'

'But I did think, after I gave you the job, that it would be nice to have someone handsome to look at, working in my garden.'

'I thought you were very handsome too. When I first met you.'

'Ah. Did you?'

'Mmmm.'

He blushed in the candlelight, as he remembered something. David noticed and smiled again, 'Oh. Did you play with yourself afterwards, Mr. Forman?'

Will gasped at the crudity, then grinned and nodded. 'Couple of times.'

'So did I. Thinking about you.'

'Really?'

'Oh yes. Thinking about what I wanted to do with you.'

His finger ran down the boy's belly and caressed him under his balls. Will gasped and said, 'Right couple of wankers, we are.'

David burst out laughing. 'Yes, I suppose we were. Not anymore.'

Will reached up and pulled him down and kissed his lips. Then he frowned. 'Tell me about your wife.'

'Eve? Well, we met at Oxford. Discovered we were both different from the other students. She fell in love with an American girl, but that ended when she went back to the States. Eve was very upset, because the girl didn't want to keep in touch. We would walk along the river for hours, just talking and talking, worried about how we were going to live with our secrets after university, and she mentioned she desperately wanted a child, but not a proper husband.'

'Had you ever been with a girl before?'

'No, never.'

'But you… did it… with her?'

'Eventually. After a lot of wine. A couple of times. Then she got pregnant so we stopped. We were both rather relieved.'

He chuckled.

'Then she died?'

David sighed heavily, 'Yes. That was dreadful. I lost a very close friend, and, of course, the baby too. I hadn't really thought about being father before I met Eve but had gotten used to the idea.'

'Do you still want to be a father?'

'No. Not now. I like children, but I don't want any of my own. Eve did, and I would have tried to be a good father for her sake, but I never really wanted kids myself. I bought Flush after the funeral. Maybe, in some way, I got him as a substitute, but it was mainly just to have some company.'

'And now you've got me too.'

David kissed him again and said, 'Yes. I have. What about girls with you? Did you ever…?'

'No. Not a one. Not even a fumble behind the village hall after a dance. We never talked about sex at home. Mum never talked about it, except to say it was dirty. When I went out with a couple of the village girls, she would warn me about sex and sin.'

'Your mother is a very sad woman. Do you miss being with her?'

'No. Not at all. She's made her bed.'

'She has, and she's lost a very sweet man as a result.'

Will sighed, touching David's face.

'I do like this. Being with you like this. Just touchin' and stuff.'
'So do I.'
He kissed Will's lips, then each nipple and slid down and kissed the top of his cock. It stiffened immediately. Coming back up, David asked, 'Do you, by any chance, want some more stuff?'
Blushing again, Will bit his bottom lip and said, 'Yes please…'

Chapter Seven

It was a beautiful July, with glorious weather. Will's predictions for the herbaceous borders had come true, and they were now a blaze of colour. The water lilies had started to flower, and they had managed to protect the shubunkins and goldfish from the heron. Apart from them losing one of the chickens to a fox, everything in the garden of Wisteria House was lovely.

Four weeks to the day from his visit to James Street, a large parcel arrived from Lobb's, containing two pairs of very fine work boots, and a pair of plaster of Paris feet. Mr Small had added a note saying, "It turned out your idea worked. Please accept these as a gift and return them to us should your friend need further footwear."

David placed the two feet on the mantlepiece in the living room, where they became a subject of much conversation, although he never explained whose they were or why they were there. Will's new boots fitted perfectly, eliminating his limp. He felt very embarrassed David had spent so much on him, but the comfort of the new boots overcame any further argument or potential distress he felt. David sent a cheque to cover the bill, enclosing a ten-pound note as a tip for the apprentice. His standing at Lobb's also rose as a result of this kind gesture.

As the summer went on, the rest of the garden took shape, reappearing from the tangle of overgrown shrubs and weeds under Will's dedicated care. David kept on writing and by the middle of August, the new book was finished, and the completed typed manuscript returned to Wisteria House by Miss Bottomely for final changes and corrections. David did these sitting outside at a round teak table in the sunshine, Flush at his feet.

He now wore shorts and an Aertex shirt most days, as the weather was extremely hot.

With all his changes and additions made, he sent the final version off to his publishers for comments and an initial print version. He would go up to town later that month and run through the galleys with the printer, to make sure no typos had slipped by his editor and proof-reader.

Whilst David's standing with the "ordinary" villagers was on the up, his reputation amongst the landed gentry was not so good. His total failure to try and seduce one of the Langham girls, or show any interest in them whatsoever, was the cause of much resentment, especially amongst the girls.

One day in late August, Florence Langham knocked on the door of Wisteria House and invited herself in for coffee. Lottie wasn't there, and David had been out in the garden, thinking about what to write in his next book. She needed an escort to a summer party at a country house just outside Gloucester and had decided David would be perfect for the task.

He listened sympathetically then said, 'My dear Florence. When I was a child, my parents, despite it being war time, dragged me to endless parties and events around Salisbury, where I had to sit and listen to boring old men and women talk about Hitler or their golf swings or their aches and pains, or I had to make small talk with dull children, equally uninterested in my presence. I made up my mind then that I would never take part in any social gathering that wasn't my thing, and a party full of hooray Henries and jolly hockey stick girls looking for husbands just isn't my thing at all. I am so sorry, but it just isn't. I am sure, and indeed hope, that you will find a charming man to escort you, but it isn't going to be me.'

'Well, I am disappointed. You seem to lock yourself up here, writing all the time. It's very dull of you, David.'

'It makes me happy.'

'Don't you want to make me happy?' she asked, flirting blatantly.

'I don't want to make you sad, but I cannot help you in this situation.'

She had left, cross that her charms had failed her this time, but determined to try again, mainly because he had looked so damn good in his white shorts and tight shirt. A man's man she thought. (If only she knew…)

David kept himself very fit, through exercises, running and lifting weights in his bedroom, when he wasn't lifting Will. His lover was naturally muscular from his daily work, but the writer had to make an extra effort to ensure he didn't become plump and soft as he sat at his desk every day.

He had just settled back in his chair by the teak table when the telephone rang. He went inside again. It was Murrey Holt. 'David! Glad I caught you. Got some news. The BBC are very interested in making a series out of *The London Connection.*'

'Really? That's wonderful. I thought you would get ITV interested, not the BBC. It's a bit bold of them, isn't it? I thought they preferred their drama to be Charlies Dickins or Emily Bronte.'

'Well, I presented them with the idea of something a little more risqué, with a bit more bite, to compete with ITV. They went for it. Not offering much for the rights, but then that's the corporation for you. Only the first book too. They want to see what the reaction is. If you agree, they want to start filming in October and send it out next March, just after Easter as their main spring drama. On Sunday nights.'

He mentioned the sum being offered. It might not be as much as the Hollywood producer offered, but it would keep David and Will in jeans and fish and chips for a good few months.

'Sounds excellent. Do they want me to do the script or do they have their own man?'

Murry said, 'They want John Coleman to do it. He did their *Wuthering Heights* and they feel he could do a good job.'

'So do I. I'm flattered. He's the best at the moment. Full English cast?'

'Oh yes, no Americans.'

'Then yes, happy to agree to all that. Thank you, Murrey. You're an excellent agent.'

'Well, I'll get my cut of course, so I'm not doing it out of love for my fellow man, you know.'

David laughed. 'Oh, I know, I know, but I still appreciate it. I'm not good at that sort of thing, as you know.'

'I do. Just keep writing. That's all I ask. How is the new book?'

'Finished. I'll be coming up in a few days to just check the galleys. They will start printing in September for an end of October release this time. They want good sales for Christmas.'

'Very good. Excellent. Are you happy with it?'

'Yes. And it is very, very racy this time.'

'Oh good. Your readers love the rude bits.'

He disconnected.

David went to find Will. He was harvesting spinach and carrots in the vegetable garden. The chickens were keeping him company, pecking the gravel path, guarded by the cockerel.

'The BBC want to film my first book.'

'Oh. That's fantastic.'

'I know. Ouch, get off, you brute…'

This was directed at one of the hens, which had pecked his bare foot.

'Could be the start of something really good.'

'You've already got something really good.'

David slipped his arms around Will's chest and kissed him. 'In more ways than one.'

'I'm all sweaty.'

'I know…' David replied, grinning.

'You're a dirty man, Mr. Manners.'

'Complaining?'

'Nope.'

'Good.'

He released his lover and said, 'I better leave you in peace. I might get carried away if I stay. Oh, I thought we could go for a run in the car on Saturday and take a trip over to Slimbridge. It might be nice to have a change of scenery.'

'The bird sanctuary? Oh, that would be great.'

Will loved all wildlife, especially birds. The only creatures he hated were the ones that ate his plants.

'Good. Okay then, see you later.'

He gave him one more kiss and went back through the arched gateway to the terrace and his work. Will watched him leave then bent down to cut off some more spinach.

He was still adjusting to being so open with his affections, in private at least, but loved the way David made him feel so wanted and welcome. He was growing more comfortable day by day, no longer a charity case but fully part of David's life and home.

David realised at three in the afternoon that the Langham girl's visit had made him forget to buy dinner. Fetching Flush's lead and taking his happy dog with him, he walked to the butcher and bought a chicken and some thick slices of bacon.

It took a while to get back home, as Flush had to greet every dog they met on the way back to Wisteria House, and pee on every tuft of grass and edge of wall they passed.

As he put the bird and the bacon in the refrigerator, Lottie came into the kitchen having finished ironing their shirts in the scullery.

'Oh, having a chicken, are you?'

'No. This is an ostrich.'

'Very funny.'

David grinned. 'Thought it would be nice. With new potatoes, carrots and spinach. All fresh from the garden.'

'You two eat very well, I must say.'

'There's plenty if you want some. Spinach that is.'

'I know. Will has already given me a big bunch of it. Flora and I will have it with poached eggs. She loves that for her tea.'

'Good.'

She was tidying up one of the drying up towels which hung over the handle on the front of the cream-coloured Aga. She hesitated then said, 'David?'

'Yes?'

'It's about Will's room.'

He leant against the kitchen sink.

'Go on.'

'There's no need for him to keep roughing up his bed. Or using the other bathroom. It's just making extra work for me.'

He was very silent, shocked that she had raised the issue and slightly confused as to how to respond.

'You noticed then?'

'I'm not daft. Anyway, my brother was a bit like you two. The one that died in the war. He was best mates with a lad from the village. They signed up together. Both of them died in France, on Gold Beach in Normandy. Never apart, even in death.'

'Oh, I didn't know.'

'Well. If it's all right with you, we can stop all that here. Saves me having to wash two sets of sheets, especially as one of them is never used…'

'Lottie Nolan. You are a very, very nice woman.'

She nodded, glad to have sorted that out. She hated doing unnecessary work.

'I know that too. Evening, David. Have a nice weekend.'

When he told Will, he went deep red.

'Oh, bloody hell. She knows?'

'Should have got those sheets a bit dirtier.'

'What do I say to her?'

'I wouldn't say anything. She might not disapprove, but I don't think she wants any details...'

Chapter Eight

It was pouring with rain when they arrived at Slimbridge the next day and the car park was empty. Bird watchers didn't seem to like the weather, which was a little surprising as they must spend a lot of time getting wet following their hobby, David thought.

They got out, noticed the sign saying dogs must be kept on leads the whole time, and fixed theirs to Flush's collar. There was no one manning the visitor's kiosk, so David put two shillings on the countertop to cover their tickets. There was a path leading to a viewpoint over the marshland and the main lake. Hundreds of ducks, geese and swans covered the surface, or stood preening their feathers on the shoreline. It was stunning.

'Morning. Didn't think we'd get any visitors today.'

It was a middle-aged man in a green jacket, who had appeared from the house. There was a little construction work going on, and a vast floor to ceiling window extension looking out over the sanctuary had recently been added to the front of the old house.

'We put our money on the counter. Anything special we should be looking out for?'

'Well, most of the fledglings are up and flying now. They'll be heading south in about a month. We have curlews, red shanks, oystercatchers. We also have a pair of corncrakes in the reeds, but you'd be a better man than I if you can see them.'

David realised this was the man himself, Peter Scott, the founder, ten years earlier, of this bird sanctuary, and son of the Antarctic explorer, Captain Robert Scott.

'Thank you, Mr Scott.'

'Ah, the price of fame. And you are?'

'David Manners.'

'Better known as Connor Lord,' said Will.

'The author? Oh, I love your books. I've got all five inside.'

58

'Not many birds in them, I'm afraid.'

'Even naturalists need a rest every now and then. And you?'

'Will Forman. I'm a gardener.'

'Strange bedfellows,' said Scott.

Will blushed in the rain, but David replied, very calmly,

'A love of birds brings many people together, doesn't it?'

'Indeed. Tell you what. Have a look around then bang on the back door and come and have a cup of tea. Perhaps you could sign my copies.'

'Happy too. Right then.'

They left Scott to dash back inside and went around the paths to the various other viewing points, where signs clearly described, with pictures, the types of birds they were seeing, their breeding habits and special features. After a very cold and wet hour, they headed back. The doorbell was hanging off the wall, hence Scott's instructions to bang on the door.

He came and opened it quickly, offering towels for them to dry their hair and a rag for Flush's paws. Then he led them to the studio living space by the big window where his wife was laying out tea for four. A painting was standing on an easel, and David's five novels, all hardback editions, were on a side table, along with a black fountain pen.

'This is Phillipa, my wife. David Manners and Will Forman. David writes under the name Connor Lord.'

'Afternoon. Dreadful day to be visiting, but the ducks love it,' she said, laughing.

David signed the books then joined the three by the table.

'It is very beautiful here. And very clear information too. We didn't see the corncrakes, I'm afraid, but most other birds. Oh, is that a Peregrine Falcon?'

They watched as the grey and white raptor, hovering over the lake, suddenly plummeted towards the flock of ducks. They all took off in panic, and the hawk disappeared into the rising cloud of birds. Then it reappeared, a mallard in its talons, and landed on the shoreline, ripping into the duck clutched in its feet.

'Yes, that's a Peregrine. Pity about the duck, of course, but that's nature for you. We seem to have created a buffet for the hawks around here.'

They watched nature taking its course for a while before Scott asked, 'Do you have a lot of birds in your garden?'

'Not really. There's an ornamental pond with goldfish and carp and we've put in water lilies to try and save them from the heron. Apart from him, not many. Any suggestions?'

'You should have a bird table and put out scraps and breadcrumbs. Bird boxes too, encourage birds to nest. And if your garden is big enough, leave a patch rough, with nettles and thistles. Butterflies love the nettles and many birds like goldfinches feed on the thistle seeds later in the season.'

Will sighed, 'I've just spent the last few months getting rid of all the nettles and thistles.'

Scott laughed.

'Most people do, but our wildlife loves them.'

David was studying the painting on the easel.

'Do you sell many?'

'Oh yes, and half the proceeds go towards the sanctuary. I sell prints too. Come and see.'

They left half an hour later with a print rolled in brown paper, a beautiful, signed copy of a painting of geese flying across the estuary at sunset.

'I'll have to go into Stroud and get this framed. It would look nice over next to the piano, I think.'

'Well, we're going in on Wednesday to see the dentist so we can take it with us then. And I'd better start planting thistles…'

*

A couple of weeks later, they were sitting by the pond having a cigarette when Will said, 'You should get a greenhouse. I could grow tomatoes and lots of other good things if we had one.'

'Where would you put it?'

'Build it on the south-facing wall of the vegetable garden. The outer wall, that is. A big lean-to one; the stone would keep the heat in at night.'

'Good idea. I wonder if Fred could build one for us.'

Will looked at him carefully then chuckled.

'I like that word "us".'

'It is us from now on. I love you so much, Will. I know I haven't said it before, but I do.'

'I only said it when I was drunk.'

'I do remember.'

His blond boyfriend looked over and touched his hand.

'I love you too, drunk or sober. I'm very happy.'

'Good to hear. Right then, I'll call Fred and see if he can pop round on Monday and take a look at the wall.'

Fred Compton was the local carpenter. He repaired windows, put up bookshelves and rehung doors, and was kept busy throughout the year, but he came round at eleven the following Monday, said a greenhouse was no problem and took some measurements.

There were some cut stones piled up at the rear of the garden, left over from previous building work. Compton looked at these, nodded wisely and said,

'I can make a low wall as a base using those, then put the greenhouse on top. My lad can do the building work. He's just finished his apprenticeship and we're going into business together. And I know they have some nice cedar planks at the timber yard that would be perfect.'

He had brought his son with him, a tubby lad of sixteen, who nodded sagely at this statement. It took a week to make the stone footings for the structure and a further week before the cedar frame went up. Fred called a contact of his in Stroud who was a glazier, and after his son, Bert, had treated the frame three times with creosote, the glass went in.

It was too late that year to plant tomatoes, but Will split loads of tubers and took cuttings from several shrubs and covered the slatted shelves with flowerpots filled with potting soil to grow his new plants.

The summer passed into autumn, the garden filled with butterflies feeding off the buddleia bushes, and bees invaded the borders, with their riot of Michaelmas daisies and snapdragons. David went to London to go through the galleys, and the new book was sent off for printing in time for its release at the end of October.

He then turned to finding an idea for his seventh Wright novel. He would take Flush and a haversack with a thermos of coffee, sandwiches and some fruit cake, and go off walking for hours, thinking about a new plot.

Ian Fleming was now on his sixth Bond book, as *Dr No* had come out to rave reviews the previous March. His more outrageous stories were very popular, but David's more ordinary war-weary hero was much loved by his faithful readers.

There was a copse of pine trees at the edge of one of the farms, with a fallen trunk that he would sit on and eat his packed lunch. It overlooked the fields and the Cotswolds Hills, a peaceful place to reflect and think.

One day, the tranquillity was shattered an English Electric Canberra, an RAF plane, which scattered the cows and boomed across the horizon, heading west. It was going so fast, and very low for a high-altitude bomber, that David wondered if it had been stolen. With that single thought, he had the idea for *The Copenhagen Case*, a tale of the theft of military hardware, stolen secret plans for supersonic jets, and a high-speed finale over the skies of the Danish capital. It would mean a visit to Copenhagen and he wondered if Will would like to go too. The lad hadn't been anywhere abroad; in fact, the furthest he had travelled from the village was Salisbury Plain for his basic training during his national service.

A month later, both clutching new passports, they landed at Kastrup Airport and took the bus into the city centre. David had bought a new box camera, and Will took loads of pictures of the streets and buildings, as David made copious notes to ensure the details were correct when he described the city in print. They had adjacent rooms in the oldest hotel in Copenhagen, Hotel D'Angleterre, but only slept in one of them and spent three days wandering around the glorious old town. Will wasn't too sure about some of the food and avoided the raw herrings. But David tried everything, just so he knew what things tasted like, again for the story to seem genuine.

They flew back to Heathrow and drove home to an ecstatic Flush. Lottie and Flora had stayed at Wisteria House while they were away, to keep it safe. Flora had never watched television before and had sat down in front of it every day after school, marvelling at the programmes. They had both spoilt Flush dreadfully, but he had still missed his masters.

When Will went to the pub on his own the next evening, he was interrogated about the trip. The main question was why his employer had taken his gardener to Denmark.

'I thought the whole point of you living there was to look after his dog when he went away,' said Grumbley, nursing his cider.

'He wanted me to carry his bloody bags, didn't he? I was like some sort of glorified valet. He was marching all over the shop, taking pictures and making notes, and I 'ad to run after him, carrying his camera and briefcase.'

This had been their agreed cover story. The farm workers nodded and thought this was a very reasonable explanation. David got some stick when he went in on Friday evening, but he just said, 'When I'm writing and thinking, I tend to forget things. Will picked up all my stuff and made sure I didn't leave anything behind. Very useful chap, he is.'

That bullet dodged, David got down to writing the new book, and life returned to normal in Wisteria House. When Lottie told David how much Flora had loved watching TV, he rented one for them, as part of her wages.

In the years ahead, "doing research" would become the code words for future trips, to the Greek islands, Nice, Rome and Southern Spain, none of which locations ever featured in a John Wright novel. If any of the people who read his books in Fletcher's Cross ever noticed this peculiar absence, they never mentioned it.

Chapter Nine

Lottie Nolan liked working at Wisteria House. Apart from being well paid for her three full days, and two half days, every week, she often went home with free vegetables, or some eggs, or a bunch of flowers from the garden, and David always thanked her at the end of each day for her hard work. The same could not be said for some of those she had worked for in the past.

It was a nice house to clean, and David was a tidy man. He never left towels on the bathroom floor, or his clothes strewn about in the bedroom. If the weather was bad, he always took Flush out through the kitchen, saving the oak floor in the living room from muddy paw prints.

Her day would start, after a cup of tea, with washing up the breakfast things, and the previous night's dinner dishes. Then she would wipe down the worktops and the kitchen table, mop the floor and head upstairs. The bathroom came next (just the one nowadays), followed by making the big brass double bed and dusting the chest of drawers and the bedside tables. There was a long, built-in wardrobe in the master bedroom, an innovation the late Mrs Roscoe had installed, with light oak doors, that held most of his clothes. It had one door that was inset with a tall mirror, which she would polish most days, then she would go down and do the lounge.

She loved this room. It was a well-proportioned space, with French windows leading to the garden, and four tall windows, two on each side of the doors, overlooking the lawn and the herbaceous borders. A stone fireplace with a large, gold-framed mirror above it was opposite the doors. On one side stood the black and white television: on the other, a brass box filled with coal.

The baby grand piano was by the door leading to the study, and on the opposite side of the room was a huge bookcase, filled with hardback novels, both classic and modern.

Arranged around an Indian carpet in front of the hearth stood a deep comfortable sofa, big enough for four people, and two armchairs, all covered with a fine William Morris print of birds and leaves. The heavy curtains, lined to keep the heat in, matched the green of that foliage. The suite had loose cushions; some in the same textile as the sofa and chairs, others green and apricot, matching other colours in the Morris print.

There were small side tables of different sizes with cut-glass ashtrays, with a silver cigarette box and matches in a silver sheath on the biggest. Two also had small table lamps, like oak candlesticks, with shades that matched the curtains.

David had few ornaments, which made it easier for Lottie to dust, but those he had were of fine quality. Two green Chinese jade three-legged toads, a carriage clock and a bronze-coloured glass fish by Lalique were the only ones on the mantlepiece originally, but they had now been joined by two white plaster of Paris feet.

There was only one photograph in the room; in fact, it was the only one in the whole house. It stood, in a silver frame, on the piano, and showed David and Eve on their wedding day. The bride was wearing a simple white dress, the groom a black suit, and they were standing by a river with an Oxford college behind them. Lottie had never been to Oxford so she didn't know which one.

They had their arms around each other and were laughing. It was one of the happiest wedding photos she had ever seen. Many couples, especially the men, looked as if they were facing execution, not the rest of their lives living happily ever after.

The dining room, between the drawing room and kitchen, was seldom used and she only cleaned it before and after a dinner party. A very fine eighteenth century walnut dining table, with two carvers and three dining chairs down each side, dominated the room. Two extra dining chairs and an extension leaf stood against one wall, and a walnut sideboard, with two Georgian candelabras, was on the other, under a mirror with a finely carved Chinese black lacquer frame.

David also owned a full canteen of Georgian silverware and a wonderful set of Wedgewood Pearlware hand painted imari plates, dishes, soup bowls and tureens, enough for a dinner party of twelve. The room had been painted a dark green, with even darker green velvet curtains.

Her final cleaning job, apart from doing the washing, was the hallway. This ran along the front of the house and was wood panelled, with a dog-legged staircase leading to the first floor. She would check the cloakroom, then hoover the carpet, dust the table that had another clock on it as well as a bowl for keys, before returning to the kitchen.

David had one item that impressed Lottie more than the silver and china. He had a second hoover, stored in a spare bedroom, which meant she didn't have to carry one upstairs.

Washdays were Mondays and Wednesday. She did the sheets and towels on a Monday, and their shirts, socks and underwear on Wednesdays. Lottie had been very impressed to find the latest top loading Hotpoint washing machine and a cylinder spin dryer installed, so she didn't have to use a mangle to extract water from the clean clothes on rainy days.

On fine summer days, she would carry the damp sheets in a wicker basket along the path by the garage and hang them out to dry in the fresh air on two lines next to Will's potting shed.

The first time she did his washing, she had asked David where his pyjamas were. She had been slightly shocked when he had told her he didn't wear any.

It was an easy house to keep clean, and her employer and his friend equally easy people to work for, and with. All in all, she thought, it had been a good day when she had got the job, and she had no complaints.

One afternoon, soon after Lottie had started, she and David were sitting in the kitchen drinking tea when she said, 'I love your home. You have some beautiful furniture and things.'

'Thank you. Most of it is from my parents, my mother in particular. Her father was a great collector of nice bits and pieces and he left her many fine things. The silver and china were his, as was the piano.'

'Very nice. But you don't have any pictures of them. Photos in silver frames. Most people like you have loads of them, all over the place. You only have one. That wedding photo on the piano.'

David had laughed, 'We weren't very good at taking photographs. In fact, the only picture I have of my mother is that drawing in my bedroom, over the chest of drawers.

And I don't have many paintings either. They had several, but most were very dark and gloomy and I sold them off, after they died. I need to buy more.'

'I like the ones you do have, especially that one of Stonehenge.'

A fine water colour of the famous stone monument hung over his bed.

'Yes, that's lovely. It was a wedding present from Eve's parents.'

'Oh. Sorry to bring that up.'

'It's ok.'

After she had had the conversation about it being okay for them to just use the one bathroom and bedroom, they had had another, about his marriage. He had explained and she had nodded thoughtfully then said,

'Well, she was a pretty girl. Very pretty.'

'She was. She was also my best friend. I miss her very much, even if I now can be with Will.'

But that day, before she knew the whole story, she asked, 'Do you keep in touch with them?'

'No. They find it too difficult. I sent a Christmas card last year and got one back, but that's been it.'

'Very sad to lose someone so young.'

'You did too. Your husband and your brother.'

'Yes.'

'Life can be very, very cruel at times.'

She had nodded, muttered "bloody Hitler" under her breath, wiped her eyes and got on with her work.

*

The coke-fired central heating went on at the end of October. Apart from the other two big houses near the village, David's home was the only one with this innovation. It also had a working fireplace in the living room, and he bought logs from one of the farmers. Will and he would lay on the big sofa at night, watching the fire, just talking, especially if there wasn't anything on the television.

Will loved TV and would have watched everything, but David was a bit more selective and was more than happy to turn it off, take his lover in his arms and just cuddle up.

Will started watching *Gardeners' World*, where gardeners talked about growing plants and vegetables. He would sit next to David and give one of two responses to the experts; "Oh, that's a good idea, I'll try that" or "That's bollocks, that is. I'd never do that." One programme that they both enjoyed started showing on the 22nd of December.

It was called *Quatermass and the Pit*, about a spaceship found buried under a London underground station. It was absolutely terrifying and began Will's love affair with science fiction and horror movies. There had been two earlier series, in '53 and '56, but this was the first he had seen. He clutched a cushion the whole time it was on, as if it would give him protection from the ghostly goings-on. David had to comfort him afterwards, in ways that totally took his mind off the scary story.

They spent their first Christmas together, a peaceful affair in a snow-covered Fletcher's Cross. They put up a tree, and David went up into the attic to find the box of Christmas decorations his parents used to hanging on their trees in Salisbury. Will watched him put a glass Father Christmas up, several glass fir cones and robins, as well as the normal red and silver balls.

'We never 'ad a tree. Mum thought it was heathen.'

Will's mother had still not come around to forgiving her son and hadn't spoken to him since she had shouted out of the bedroom window that terrible day. She hadn't spoken to David either. If she was serving in the bakery when he went in, she would go out to the back and the baker himself had to come through.

She finally joined the Catholic faith, taking the bus three times a week to Combe Weston's Catholic church, where she confessed her sins with growing enthusiasm. The Reverend Fallow seemed relieved by this departure from his flock as, as he said to David one Sunday, 'she would test the patience of a saint, and those Romans have a lot more of them than we do…'

In January, the garden started showing its hidden secrets. A carpet of snowdrops appeared around the pond, followed by hosts of daffodils. Frogs mated and filled the still water with their spawn. The apple trees blossomed, as did the new pears trees in the vegetable garden.

They had followed Peter Scott's advice and put up several bird boxes, in the orchard, one the willow trees and a couple attached to the house and the old stables. They were now visited by various birds assessing their suitability for future homes, and blue tits and a pair of flycatchers settled down to some serious nest building during March. They had also put up a bird table outside the kitchen window and, as promised, it brought in many new birds to the garden. A pair of collared doves arrived, sparrows by the dozen, and some long-tailed tits, with their pink and black feathers. They still had to be very discreet about their relationship. They went to church separately, and the pub too. David would stand at the bar and chat to Barney Young, and Will would sit with his old mates at one of the round tables.

Flush started to get confused as to which of them he should sit with and ended up in the middle of the floor most times they were in together.

Whenever the vicar popped around, which he did once a month for a sherry and a chat with David, Will would have to sit in the kitchen, reading the Daily Mirror. Afterwards he'd grumble a bit about being left out, before a hand started unbuttoning his shirt and another hand undie his fly and his mind was soon on other, more pleasant, matters.

On Easter Sunday, March the 29th 1959, the first episode of the televised version of David's book, *The London Connection*, was aired by the BBC. As it had been a very wet weekend, most people were at home and in need of cheering up. The first episode was well done, exciting and hugely popular.

The BBC drama department called Murrey Holt the following week, asking to buy the rights for the rest of the novels. This time the price they paid was much, much higher, and his commission more than enough for him to help him finally purchase his first Rolls Royce.

The occupants of Wisteria House had watched it together, then gone to bed after Flush's final trot around the garden. After some serious carnal pleasure, they lay together, just enjoying the touch and feel of each
other's body. They were too excited by the programme to sleep. For David, finally seeing his creation on the small screen had been an extraordinary moment, and the BBC had done a very good job bringing his spy to life.

Will, stroking David's chest, said, 'Can I ask you somethin'?'

'Of course.'

'If I hadn't got drunk after I'd been arrested and told you I loved you, would you have said anything? To me, that is, about how you felt?'

David kissed the top of his head and sighed, 'Probably not. I wanted to, before, but didn't know if you were interested. If I'd said something and you were not like me, then it would have been awful. You would have stopped working here and I would have had to sell up and leave the village. You remember how the locals reacted to even the suspicion of your liking men. If you'd gone to the pub and said I'd tried to get into your pants, it would have been the end for me. Maybe even arrested.'

'That's so sad. And unfair. There must be hundreds of men like us who can't say how they feel.'

'Thousands, maybe even more.'

Will sighed, nuzzling against his chin. 'I wouldn't have said it sober.'

'Then I am very glad you got drunk.'

'Me too. Do you ever get drunk?'

'Not since my Oxford days. I saw many men, and women, drink far too
much and they ended up doing stupid things... or saying them.'

His lover looked at him steadily for a few moments then shook his head in disbelief. David winked at him, grinning, as Will finally said, 'It's good thing one of us is stupid then, ain't it?'

'Yes. Very good.'

Will rolled over so he was laying on top of him. He sat up, put his hands on David's shoulders, leant forward and kissed him. He sat there, straddling David's chest, grinning, then kissed him again, very slowly before whispering in his ear, 'You're the best thing that's ever happened to me.'

'I feel the same about you.'

'Better than the books?'

'Yes.'

'And the television series?'

'Much better than that even.'

'Good.'

He kissed him again. He was pushing back on something long and hard that was sticking up between his arse cheeks. He lifted his bum up and it flopped between them. 'You're always ready for action, aren't you?'

'With you, yes. Why? Do you want me to ravish you some more?'
That word had become a joke between them due to an old Hollywood film set in Napoleonic France they had watched on TV one evening. One of the characters had said in one scene that "He ravished me in the bedroom, the beast".
Will had turned and asked, "What does ravished mean?"
'To make love to someone roughly and passionately.'
'Oh, like you do sometimes…'
'Exactly.'
When David had asked Will if he wanted some sex later that night, his lover had thrown himself on his back on the bed, raised his legs and said, 'Go on, ravish me, you beast!'
David had laughed so much he had almost been unable to perform. Almost. They now used it every now and then and, that night, Will thought about the offer, kissed him and whispered, 'Oh, go on then...'

*

On the second of April 1959, David stood outside the house on the green looking at the village. It was exactly one year since he had moved here. He had arrived alone, apart from his faithful dog, unsure of what sort of reception he would get and wondering if he would fit in. Now he felt part of the local community, had a lover living under his roof, a career progressing very well indeed, and a home he adored. All in all, things had gone well that first twelve months. Who knew what the future had in store for them?

1968

Chapter Ten

'Nervous, Davey?'
'Of course, I'm bloody nervous. If he hates it, we're up shit creek without a paddle, love.'
Will handed him the Sunday Times Literary Supplement and David looked at the front page. His latest novel, *The Endless Storm*, was the only novel mentioned; it was the major fiction review that week.
He read through it quickly, then smiled, sighed with relief and put the paper down on the kitchen table.
'Oh, thank God, he likes it. He likes it a lot. It's a stunningly good review. Halleluiah!!'
'Well, come on then, read it to me.'
His partner sat opposite, buttering a thick piece of toast. They had been together for over ten years now and had never been happier. David took a sip of tea, cleared his throat and started.
"When Connor Lord killed off his secret agent, John Wright, three years ago at the end of his tenth adventure, publishers and readers alike were stunned. Never before had a writer, at the height of his popularity, ended a series so abruptly and so finally. The big question afterwards was, what sort of novel would Connor Lord write next? He had made it clear, in a rare interview shortly after the publication of Wright's last story, 'The Minsk Machine', that he was finished with secret agents and wanted to find new themes to write about.
He has turned, it seems, to the world around us and the environment in particular. Quoting Rachel Carson, whose book, Silent Spring, raised the issue of Man's impact on the natural world, Lord has taken heed of her warnings and created a modern science fiction epic, set in the near future, where Mother Nature, through epidemics, insect plagues, rising sea levels, and prolonged periods of terrible weather, including droughts and storms, has brought

mankind to its knees.

The book is meticulously researched and tells the stories of several different groups and families across the globe as they struggle with the changing world around them. These include fishermen in Indonesia unable to catch fish; farmers in Australia seeing their cattle died of thirst, or their land consumed by wildfires; African countries decimated by locusts; coastal erosion and massive high tides sweeping English seaside resorts from the map, and floods in London, killing hundreds of thousands.

Lord's gift is making the reader care about the individuals and their struggles. He was always good at creating a feeling of place, but in "The Endless Storm", he interweaves several different storylines without ever losing pace or the tension that builds slowly, chapter by chapter.

It is a bleak story, but fully of humanity, courage, ingenuity, political incompetence and downright madness. Some of the situations and imagery of destruction are so well written, so powerful, that I had to put the book down several times simply to catch my breath. This might be fiction, but it is storytelling of the highest order, and should be required reading for any government concerned about the future welfare of its citizens. It is an absolute triumph, and a welcome return of one of Britain's favourite authors."

'Wow, that's great. Oh, well done, Davey.'

David sat back and grinned. It was a huge relief, and he knew Norman Hilton, his publisher, would be equally happy this morning. He went to fetch his cigarettes from the living room, and as he came back through the hall, as if on cue, the telephone rang. David, lifting the receiver, said,

'Fletcher's Cross 780.'

'David? Norman. Have you seen…'?

'Just read it.'

'Christ, he loved it. He bloody loved it. Congratulations, my friend.'

'Thank you for taking the risk, Norman. And waiting for me to finish it properly.'

'It was well worth the wait. Oh, this is wonderful. Are you coming up to town any time soon? I must buy you lunch.'

'I'll let you know. What was the initial print run?'

'Fifty thousand. I think we'll need more after that.'

'Excellent. Well, have a nice Sunday and give my love to Gloria.'

Hilton, who had lost his wife in 1964 to cancer, had recently married his young assistant, a beautiful blond woman who had transformed his company with modern advertising and TV tie-ins, which had boosted his back list and revived the careers of some of his less successful authors.

David put the receiver down, went back into the kitchen and pulled Will to his feet. Slipping his arms around him, he kissed him and then pulled back a little, smiling.

'I wish we didn't have to go to church. I could ravish you right here and now, over the kitchen table.'

Will glanced at the clock. It was ten thirty. Even if they were quick, they couldn't manage such an act of mutually desirable depravity and still get to the service in time.

'You can have me afterwards. How about that?'

'That's a date. Come on, let's have a cigarette, then clean up and get going. Flush takes so long to walk there now.'

Poor old Flush was over eleven years old and suffering from arthritis. His snowy white muzzle and wobbly gait were constant reminders that his time was drawing to an end.

They had discussed getting a puppy, but David felt it would be unfair to the old dog. "Let him enjoy his last few months in peace, without being tugged about by a younger animal. Anyway, I'd never replace you while you were still around, so why should I do such a thing to poor old Flush?"

They walked together, letting the dog take his time, sniffing the verges along the green. It had been a while since they had felt the need to arrive at the church separately. As far as the villagers were concerned, they were just David and Will now, close friends who shared Wisteria House. No other words were used to describe their relationship.

'Hear, have you worked out that bit about the baby yet? At the end?'

They had been to see *2001, a space odyssey* the night before in Stroud and, while Will had understood the mysterious monolith, and loved all the spaceships, the star baby at the end had confused him totally.

'Not really. I hear Arthur Clarke is going to publish his script in book form. Maybe he'll explain what he meant a bit better in print.'

'I thought it were by that Stanley Cruikshank.'

'Kubrick. Stanley Kubrick.'

'Oh, right. Well then, I thought it were by him.'

'He was the director.'

'Bloody good spaceships though. So real.'

'It was wonderful. I loved the way he used music too. Bloody hell, what's going on here?'

They had reached the churchyard gate. There was a group of young girls and boys standing outside St. Stephen's, and Fallow, in his black cassock and snowy white surplice, was looking rather surprised at these sudden additions to his flock. The kids were wearing bellbottom jeans, tied-dyed T-shirts, beads and all had long hair; it was a mini hippy invasion.

'Morning, James. What going on?'

'Flora Nolan's back in the village. Visiting her mother.'

'Ah, yes, Lottie said she might be doing that. What's this then? Her fan club?'

'Yes, I've told them if they want autographs they will have to wait outside. Hello Flush, old boy. Ah, that's it, take a seat by the bench. He can come in, you know.'

'He'll be okay. This lot won't bother him.'

Some of the kids were patting the Labrador but were keeping an eye on the church door just in case Flo appeared. The old dog settled with a pained sigh on the grass by the bench. David chuckled and said, 'It's a nice day and he likes the warm sunshine. Soothes his aching muscles.'

They had arrived at the church with three minutes to spare. David slipped the lead through the back of the bench and they went inside, sitting together in the back row. Since Will's mother, who had died the previous year, had gone Catholic, they had done this without comment or criticism from their fellow worshipers. Mrs Forman had not forgiven Will, who had never returned home, and their relationship had gone from bad to worse, not even improving slightly when she went into hospital for the last time. Cancer too, had taken her away, and Will still felt slightly guilty feeling that her departure from this life had made his much easier.

Flora Nolan, or Flo Nolan as she was known publicly, was sitting next to Lottie at the front. She was currently number one in the record charts with her latest single, *Take good care of my heart*. She had appeared on *Top of the Pops* that Thursday, the new music programme on the BBC.

She was that rare thing; a female singer songwriter now, with a deep raunchy voice that was wowing the public. She came back to the village every now and then and, that day, was trying not to disrupt the proceedings. That was difficult considering she stood out as a glorious riot of colour, beating the stained-glass windows into a cocked hat.

She had her black hair styled in a bob cut, heavy eye make-up and was wearing a shiny plastic Mary Quant raincoat of a deep purple colour, over an orange mini dress. The people around her looked shocked at this alien figure amongst the tweed and grey suits. Every now and then there were screams from outside, as her fans got overwhelmed with excitement as they waited for her to finish praying and come and talk to them.

With the service over, David rescued Flush and they stood to one side by some gravestones as Flo made her appearance to wild adulation from the gang of young people waiting for her. They crowded around her and she signed record sleeves and autograph books, happily chatting away.

Lottie came up to David and Will and said, 'You wouldn't credit it, would you? So shy at school and now look at her.'

'She's a star, Lottie. And a bloody good singer.'

'Oh, I know, but these kids. Some were hanging around outside the cottage all last night, singing her songs. The neighbours started complaining.'

'It will pass. All things do.'

Lottie snorted, unhappy at the unwanted attention. Then she remembered the new book and asked, 'Here, how was the review? I don't get the Times so you'll have to tell me. It wasn't in the News of the World.'

Will laughed and said, 'What do you think, Lottie? David's brilliant and they loved it.'

'Oh, well done. Super. I know you were a bit worried. Are they goin' to leave us in peace to have our Sunday dinner, love?'

Flo had joined them. 'Hi David, hi Will. Yeah, I've given them what they wanted.'

'Love the new song, Flo,' said Will.

'Ta, straight in at number one. Wonders will never cease, eh?'

'Is it true you going to America at last?' asked David.

'Yeah. Going off on tour with The English Hooligans in three weeks, can you believe that? Forty concerts in fifteen weeks.'

That was a band, more like The Rolling Stones than The Beatles, who were very popular with wilder rock music lovers.

'Take good care of that voice of yours, won't you?' said Will

Flo laughed, getting the reference. 'I will, babe. Come on, mum, I want your Yorkshire puddings and roast beef.'

They went off, arm in arm, as David and Will walked back to their home.

'Fancy our Flo touring the USA.'

'Did she look a bit high to you?' David asked.

'Maybe. So many of those rock stars do drugs. Or so the papers say...'

Chapter Eleven

The sixties had been a period of change, in Britain, America and at Wisteria House. The decade had started with an ancient conservative British Prime Minister and a thrusting, young American President, and now there was an ancient President sitting in the White House and a charismatic Labour PM, Harold Wilson, in Number 10. The events in Dallas in 1963, and the removal of the Conservatives from power, had transformed politics.

Socially, Britain was becoming a brighter, hippier place. The Beatles and their Mersey sound had given the world pop music, and the psychedelic drug culture that had styled the younger generation, Flo Nolan being a classic example.

The Beatles and the Stones represented two sides of this and the formers' transformation from guys in black suits to long-haired, bell-bottomed seekers of truth saying "Peace, man" in most interviews led many teenagers to copy them; some even embraced LSD and marihuana. David and Will had all their albums at home and loved their new music. The *White* album was really good, they felt, and Will especially loved the song, *Blackbird*. They often heard one in the garden singing at the dead of night when they let Flush out.

In London, Carnaby Street was the focus of fashion. The post war drabness had given way to bright colours, wide ties, droopy long moustaches and much confusion amongst the older generation. David and Will bought some of these new items, but stayed pretty traditional, with straight legged Levi 501 jeans and denim shirts, although Will started wearing Henley collarless two button long-sleeved shirts, as those drove David wild with desire.

The author was rich now and had realised he was always going to be the breadwinner in their little family.

Apart from his book sales, he had inherited money from both his parents when they had died, which had helped finance his life, and the purchase of Wisteria House. He focused on making sure he retained as much of his wealth as possible. He invested in shares, art and property. He purchased a one-bedroomed flat in Florin Court, overlooking Charterhouse Square in London, so they didn't have to book two hotel rooms wherever they went up, and he also bought the cottage to the left of Wisteria House as an investment in 1962. He rented this out to one of the local schoolteachers and his family.

His art collection started when a country house outside Stroud held a house sale during a snowstorm, and he managed to buy a Constable, a Stubbs and one of Van Gough's sunflower paintings all for under a thousand pounds. He also bought from London galleries; two paintings by an up-and-coming British artist called David Hockney and he had invested in two American artists as well. He had a huge canvas by Roy Lichtenstein which now hung over their bed, and an Andy Warhol print of Marilyn Monroe on the wall by the piano. Stonehenge had been moved onto the upstairs landing, and the Peter Scott print hung over the table in the hall.

The photograph of his wedding day had been put away, and they now had several photographs of the two of them around the house. They had two favourites. One of David, sitting on the garden bench, "talking" to Flush by the pond, and one of Will, wearing just jeans, leaning against the wall of the house, arms folded across his chest, looking very moody and sexy.

During the worse winter in English living memory in 1965, the old couple who lived in the cottage on the right-hand side froze to death, and he bought that cottage too, after selling the little house in Jericho he had lived in in Oxford with Eve. He wanted his assets a little closer to home.

That cold winter caused a lot of damage, as external water pipes froze throughout the village and the spring saw plumbers and builders working hard to move water supplies inside. The arrival of gas in the village saw David replace the old coke-fired boiler for one using North Sea gas to provide their heat and hot water.

The new youth culture was also the trigger for him to kill off John Wright, secret agent. His readers had grown old alongside his hero and, even if he had worldwide sales in the millions,

David knew one of the problems of being a popular writer was staying popular. Young people didn't want to read about the post-war period; they were interested in other things, like love and nature. It was during another visit to Slimbridge that he heard Peter Scott telling a group of children about the loss of wildlife due to mankind's activities that first made him think of the impact modern life might be having on the world.

It had taken him nearly three years to research and write *The Endless Storm*, and it had been a huge risk for his publisher. Despite his record of high sales, it was new territory for both of them, an environmental thriller.

During the time he was writing, they didn't go up to London so much, so he rented out the flat to an American diplomat. They sometimes had day trips, but tended to come home afterwards, not stay overnight.

When it came to entertainment, they loved watching television and went to see a lot of films in Stroud. Will loved *Doctor Who*, which had started five years earlier, even if he became very confused when the first Doctor transformed into the second. The idea of a man who travelled about in a spaceship disguised as an old police box was amazing, he thought. The police box in Fletcher's Cross had just been removed. "It's gone and dematerialised" said the gardener, the word for when the Doctor's police box disappeared on another flight, when they noticed it had been taken away. David thought the idea of regeneration, the term used when William Hartnell had been replaced by Patrick Troughton in the leading role, was simply brilliant.

They also both loved *The Avengers*, with Patrick McNee as the suave John Steed and the glorious Emma Peel, played by Diana Rigg. They watched the news every night, and a couple of the BBC's comedy programmes, *Hancock's Half Hour* and *The Likely Lads*. David also enjoyed the new satirical shows such *That was the week that was* and *The Frost report*, both of which poked fun at politics. They never watched the new soap operas that ITV broadcast, but Lottie avidly followed *Coronation Street* and *Crossroads* and would tell them about certain characters and their antics. "That Ena Sharples. Got a mouth on her like a navvy."

They bought a lot of records (David also bought Will a portable radio for his potting shed so he could listen to music while he worked), were keen on Radio One when it started, and often listened to music in the evening.

David still played the piano, adding to his repertoire by playing pop songs as well as the classics.

They had had to be very careful and discreet throughout the sixties, until homosexuality became legal in '67, never expressing their affection in public. This meant that their home was very important to them both. They loved being together and didn't need other people to make them happy. They never cheated on each other; indeed, their mutual sexual attraction increased, and sex, both gentle or hot and sweaty, was always an acceptable way of spending an evening.

Early on in their relationship, David had made it very clear that, as far as he was concerned, his money was their money. Will must never feel beholden to him. The second thing they agreed on was that they would always try and say "yes" to the other's suggestions. He had said, 'Most couples fall out over stupid disagreements over what to eat or which film to go and see. Let's try saying "yes" all the time. It will balance out and we might avoid having arguments.' It had worked.

If he suggested lamb chops for dinner, Will said yes. If Will asked if they could have sausages and mash the next day, David said yes. It became a habit, and neither of them had any reason to regret it. But the secret to their happiness wasn't their lust and love or the lack of conflict. It was the simple fact that they liked each other and had become best friends as well. They enjoyed each other's company over everything, and everyone, else.

*

The whole of Fletcher's Cross had been thrilled when Flo Nolan had been noticed by a record producer when she was singing one of her own songs at a local talent show in Stroud. She had always made their Sunday services special with her singing, but now she was a major star, heading for world-wide success. Even the arrival of her fans that weekend hadn't put people off her.

But the village had begun to change in other ways than being a place with a bunch of young hippies hanging around. As the two friends walked back from the church that Sunday in early October, some of those changes became apparent again.

'Christ, it's getting more and more busy. Even now in the autumn.' As they had walked along the village green, David had noticed the additional cars near the Fish and Trumpet; more people "having a nice day out".

Two years before, as visitor numbers in the Cotswolds started to climb, Barney Young had bought the cottage next to the pub and knocked through, extending the bar and adding a function room at the back, for weddings and special birthday parties. He now served sandwiches, scotch eggs and pickled onions and sold masses of packets of crisps to the passing trade every weekend.

In high summer, the pub had become popular with day trippers, who sat on the low wall outside and tossed their empty crisp bags on the pavement, to the general annoyance of the neighbours. Fletcher's Cross was much the same otherwise, with just a few subtle differences, and one major one.

There were many more TV aerials now, as everyone had television. The buses only came through four times a day instead of six, as more people had their own car. One of the dairy farms had gone bankrupt. Some of the fields had been taken over by the neighbouring farms, but the farmhouse and twenty acres had been bought by Peter Franklin, the local builder, who had knocked the old place down and built an estate of fifty new houses and bungalows.

Between this estate and the old village, he had also built a Spar supermarket and a new pub, the Fletcher's Arms. None of the original locals went there, and it was struggling to survive. The main reason being that Franklin had built nice homes for a change, in stone, and the sort of people who had bought them didn't want to play slot machines in a bar with loud pop music playing all the time.

The new pub's main customers came at the weekend and were the motorbike brigade that roared along the country lanes at high speed. The new homeowners hated them.

The arrival of the Spar had not greatly affected the three village shops but had saved the locals from their weekly bus trips to Stroud for many items unavailable in the greengrocers, the butcher and the bakery.

The new store also provided much needed local work, so was accepted and put up with, rather than welcomed and loved.

The newcomers mainly commuted from Fletcher's Cross to Stroud and Gloucester and wanted a peaceful time in the countryside; as a result of the influx of new families, the local council were currently building an extension to the school, to accommodate the extra children.

Franklin's development had inspired David's other big purchase during this period. He had become worried about future new housing in the village impacting their home.

One night in the pub, he had been chatting to the farmer, Keith Robson, who owned all the land behind Wisteria House. He was complaining about bank charges for his overdraft and mortgage. David suggested he buy the field directly behind their home, a hundred acres of grazing land. He would rent it back to the farmer at a peppercorn rent for twenty-five years and give the man enough cash to clear most of his debts.

Robson had leapt at the chance, so now Wisteria House was surrounded by their own property, and safe from the Franklins of this world. No one could buy the field and redevelop it.

Lottie still worked for David and Will but only three days a week. Flo had given her money, happy to share her success, and she really didn't need to work, but she would have got bored just sitting at home and enjoyed the company of the two men.

At Wisteria House, there had been a few other changes as well. The Rover 90 had been replaced by a maroon Jensen Interceptor, which combined solid bulk with speed. A lava lamp now moved silently at night on one of the side tables, and the television was the latest colour model. A stereo system, designed for Bang and Olufsen by David Lewis, stood on a purpose-built shelf, with two loudspeakers placed in corners of the room.

Will and David collected many albums, which sounded amazing on the Danish system.

Returning home after the church service, David checked the chicken roasting in the slow oven and slid two saucepans with the potatoes and carrots onto the two Aga hotplates. Once the potatoes were boiling, he drained them and dropped them in a roasting tin, with duck fat spitting in the heat, and put them in the top hot oven to crisp up. He made the gravy and dished up the bird and the carrots as Will set the kitchen table for just the two of them.

They had decided to be friend-free this weekend, just in case the book review had been a disaster. Sitting down once everything had been served, Will raised his glass of white wine and said, 'To *The Endless Storm*.'

'Hear, hear. It is a huge relief, to say the least.'

'Do you think the BBC will want to film it?'

They were currently producing the final Wright novel for showing the following year. The series had proved very popular, even if Bond, James Bond, was box office gold in the cinemas.

'Might be a bit too expensive for them. Loads of special effects if they want to do it well. It might be time to look towards Hollywood.'

'Gosh. Would we have to move there?'

'God no. I'd hate that. Wouldn't you?'

They had been to Los Angeles the year before on holiday. While they had been enchanted by the city and the flashy cars, colours and style, they had both agreed they missed the rain and the garden. Palm trees were no substitute for Will's vegetable garden and the herbaceous borders.

David went on, 'I'd be happy to go again for another holiday and next time we should see San Francisco too, but live there? No. No way.'

'Good. I'd hate it.'

'You could grow avocados.'

'I'm trying to grow them in the greenhouse.'

The first greenhouse had been replaced with an even larger structure that spring, and now they had tomatoes, a grape vine, peppers and chillies growing in there.

'How's that going?'

Will grunted and focused on his chicken. 'This is good.'

'Thank you.'

As David put his knife and fork down, main course finished, the new telephone extension, a Trimphone in pale green, hanging on the kitchen wall, rang. It was Murrey Holt.

'David! Tried you earlier but you were out.'

'We went to church.'

'Oh yes, I forgot you like doing that. Just wanted to say great review. Well done.'

'Thank you.'

'Listen. I've just had a call from the manager of Harrod's book department. He wanted to know if you would do a book signing on Thursday the 30th. He's bought two thousand copies.'

'God, he doesn't expect me to sign that many, does he?'

'Probably not but it's a great way to meet the public. What do you say?'

David was studying the calendar on the wall, where all their appointments were noted down, under a nice colour picture of this month's flower, an orange chrysanthemum.

'I'm free. So's Will. Um, what time?'

'Two until six. They open late on Thursdays.'

'Right then. Will you be there?'

'I'll pop by around three. Just go to the book department at one-thirty and ask for Richard Short. He'll take care of you.'

David untwisted the cord, hung the receiver and sat down.

'Fancy a trip to London the week after next?'

'Great. What's it for?'

'A book signing. At Harrod's.'

'Ooh, aren't you posh.'

David laughed. 'You knew that all along.'

'Yeah, I'm just your bit of rough trade.'

His lover leant forward and kissed him.

'You have always been much, much more than that. Anyway, we could stay the night and book tickets for *Cabaret* while Judy Dench is still in it.'

Even if it had opened in February that year, it was still the hottest ticket in the West End. Will frowned and asked, 'Could we get any? It is so popular.'

'Murrey could, I'm sure.'

'Well, yes then. Excellent.'

David got up and put the plates on the draining board. Then he opened the fridge and got out the two crème caramels he had made before breakfast, whilst waiting for the papers to arrive.

'Oh, my favourite. Great lunch. I hope you haven't forgotten you're going to ravish me later this afternoon.'

'Certainly not. Highlight of my day that will be.'

'Better than the review?'

David grinned, 'Of course.'

'I'm sure that's bullshit but I appreciate you saying that.'

'We'll take Flush out for a little stroll around the garden then lock the doors, batten down the hatches and I'll pleasure you for as long as you want.'

Will grinned happily, 'That could be a very long time…'

Chapter Twelve

Great joy can often be followed by great sadness, and so it was the next morning, when David went down to make tea and discovered Flush had died in the night. He lay, curled up as if asleep, on his basket by the back door, stiff and cold.

They carried him down the garden and buried him under one of the trees by the pond. There would be no headstone, but Will planted some daffodil bulbs over the body later so their beloved dog's final resting place would always be remembered.

David was very sad for the next few days. He focused on writing, but didn't go for any walks, as it seemed strange to do that without his old friend alongside.

Murrey called him the next Monday to ask if he could go to London earlier on Thursday and record an interview for BBC Radio Two's Woman's Hour in the morning, talking about the new book. He agreed, without much enthusiasm. He hated talking about his novels and had only given one interview before. On the day, with the interview done, they went to Harrod's, where David was given coffee then put behind a wide table, piled high with copies of *The Endless Storm*. There was a huge crowd of people waiting behind a sign saying, "Queue here".

Will whispered, 'I'll have a look round the store. Good luck,' and wandered off. After visiting the menswear department and the food hall, he found himself in the pets' department. This was very famous in London, as they sold all manner of beasts, apart from the usual rabbits, cats and dogs. That day they had two spider monkeys, a lion cub, several grey parrots and a boa constrictor. Will stared at the lion cub in disbelief; fancy buying one of them, he thought. That would shake up Fletcher's Cross no end.

He finally moved along to look at the puppies. There were three West Highland terriers, two black miniature dachshunds and a single golden retriever. This was a boy, all floppy eared and loose limbed, which sat and yapped at him through the glass wall. He asked to see it and the assistant handed it over. The puppy squirmed happily in his arms, licking his face, and Will fell in love for the second time in his life.

'How much is he?' he asked the assistant.

'He's got a very good pedigree. Ninety pounds. He's three and a half months old.'

Will winced at the price. He was now on twenty-five pounds a week, just above average, so the puppy represented nearly four weeks' work. But then he didn't have to pay anything towards the housekeeping, or any of the other bills, so he was actually doing really well.

'I'll have him but I can't take him with me now. We're staying at a hotel and have tickets to the theatre tonight. Can I collect him in the morning, around nine-thirty?'

'Of course, sir.'

'Could you meet me with him by the main entrance on Brompton Road so the taxi doesn't have to wait too long?'

The girl grinned. She had had far more bizarre requests. 'Of course, sir. Do you need any things for him?'

'We have a dog's bed at home, and bowls and things. But a new collar and lead would be good.'

After choosing those, he sat on the floor playing with the puppy for an hour before he wrote a cheque for the final total of ninety-four pounds, three shillings and sixpence, gave the puppy a final pat, and went off again. He wouldn't tell David about his purchase. It would be a surprise in the morning. They had talked about getting a new dog, but David had been reluctant, as he was still in mourning over Flush.

The author ended up signing over seven hundred copies of his book and, with a very sore wrist, they left Harrod's at ten past six, had a quick bite to eat at Rules, then slipped into their seats at the Palace Theatre for a stunningly good performance of *Cabaret*. Murrey had got them the tickets as David had predicted. Judy Dench lived up to her reviews with a gloriously rude interpretation of Sally Bowles.

After breakfast in the morning at Brown's Hotel, where they were staying, they grabbed a black cab for the station, but as they set off, Will said, 'I forgot my wallet in Harrod's. I called them and they have it for me. We can stop on the way.'

'God, that was lucky.'

'Mmmm.'

The black cab pulled up outside the main entrance of the famous Knightsbridge store; Will ran in, watched intently by the doorman. People simply didn't run into Harrod's. But he smiled when Will came out moments later with a very bouncy golden retriever puppy. He got into the back of the cab, grinning, and said, 'To Paddington Station, cabby.'

David was speechless. The puppy was all over him, licking and yapping. Will said, 'Couldn't resist him. Do you mind?'

'He's wonderful. Great idea. Is this what you were up to when I was signing the books?'

'Yep. It was him or a lion cub. Thought Lottie might find that a bit too much, even for her.'

'I think you're right. What shall we call him?'

'Well, we ain't calling him anything to do with plumbing and toilets.'

'He was called Flush because he used to flush birds out when we went for walks.'

'Oh. Well, I think we should call 'im Henry. Henry the First.'

'Hmmm. Okay then. Henry it is.'

The journey back to Stroud was a bit chaotic. A cute puppy on a train always generated a lot of attention, and there was a group of school children who found him irresistible and kept coming to look at him as he lay at their feet, chewing his new lead.

Lottie, taking one look at the puppy when they arrived home, said, 'I hope he's house trained,' before patting his head and getting on with the dusting. He was very well behaved, didn't damage things and never made a mess inside the house. However, Henry did have two bad habits.

The first was his tendency to take things to his bed. If a sock or a glove went missing, they could be found there, tucked under his blanket. Dusters, tea towels and slippers were also favourites, and it became routine for the three of them to search his bed if anything had disappeared. His other bad habit was digging holes. This became apparent the following spring when the ground unfroze.

If he was with Will in the garden, he would join in if Will was turning the soil, scattering dirt and stones over the lawn.

Will still worked as David's gardener, and was fully employed by Wisteria House Limited, the company set up to handle David's royalties. In this way, he became tax-deductible. They never discussed this; Will still wasn't a "kept" man and worked hard for his money.

The Endless Storm remained top of the best-seller lists on both sides of the Atlantic through November and December and for most of 1969.

Murrey had discussions with several Hollywood agents interested in buying the film rights, but the sheer cost of any production stopped them moving forward. Even if special effects were becoming more realistic, they were hugely expensive. As David didn't need the money, he was happy to wait for them to develop if it meant the book could be filmed in the best possible way. He was now writing his next novel, set in England, about the effects of the accidental release of a new toxic compound into the water supply.

His study was filled with books about pollution, and he carried on correspondences with several bio-chemistry professors at Oxford and Cambridge, trying to get the details right. He also made several trips to Manchester, which was to be the setting for the novel, and visited the sewage works for a day to try and understand what happened to water and waste in a major city. Will, unused to being alone at the house, took Henry to bed with him, and they had a few problems afterwards as the puppy would creep upstairs during the night and slip under the eiderdown, to wake them with his very wet and busy tongue in the morning.

*

In March, Flo Nolan returned to Fletcher's Cross with some news. She was five months pregnant. It had happened on tour, but she didn't know who the father was. A rather shocked Lottie sat in David and Will's kitchen, telling them all about it.

'It was a very wild tour, she says. Loads of parties and drugs. Flo says the father could be any one of a number of boys and men. I feel so ashamed.'

'Don't be, love,' said David, 'these things happen. That's show business for you. Will she stay with you until the baby is born?'

'Yes. It's due at the end of July, but then she's meant to be going off on tour again in September. She wants to leave the baby with me.'

Lottie was nearly fifty now, so looking after a small baby full-time was not exactly top of her list of things to do with her life. She had always wanted to be a grandmother, but this was not the way she had expected it to happen.

'Do you want that?' asked Will.

'Better with me than her dragging the baby around the world with all them drugs about. But how can I keep working here?'

David frowned and said, 'Do you want to?'

'Yes. I love working with you boys.'

'Then bring the baby with you. I can keep an eye on it when you're around the house. It can lay in its basket by my desk.'

Lottie looked at David as if he was mad.

'You really don't know much about babies, do you, love?'

Flo stayed at Lottie's cottage throughout the spring and early summer, getting bigger by the day, writing new songs. She had a steady flow of visitors from her record company, as well as other singers and pop stars and some very shady characters Lottie was never introduced to. It soon became clear she had a serious drug problem, and Lottie became increasingly worried about the effect they might have on her unborn child.

She asked David if he would have a word with her and a very large Flo came with her mother one day in late June and, as Lottie cleaned the house, she sat in the garden with the author, drinking tea. He wasn't sure how to start, but Flo noticed his awkwardness and said, 'Did mum want you to talk to me about something?'

'Yes.'

'Drugs?'

'Yes.'

'Oh. She worries too much.'

'She thinks you smoke too much. And take too many other things as well.'

'I've got it under control.'

'Why do you need them?'

Her hair had grown longer since she had come home, and now she pulled a strand into her mouth and chewed on it, saying, 'I don't. I like them. Got used to them on tour.'

She hesitated then went on, 'I get stage fright, David. It started before I went to America, and it got worse there. I want to sing so well, but I get scared I might not. Taking a spliff before, or a snort of coke… it helped. Helps.'

'Have you thought about talking to someone about it? The stage fright?'

'Like who?'

'Well, a psychiatrist, perhaps.'

'I ain't mad.'

'I know that, you daft bint. You don't have to be mad to see someone like that. They might be able to help you learn to relax more before a show.'

'They are called concerts now, David,' she said, giggling.

'Oh, right, yes.'

She grinned from under her long fringe of hair, then bit her lip. 'Maybe after the baby is born. Can't cope with any more doctors or midwives or medical people right now. And I don't do that many drugs.'

He wasn't sure she was telling the truth but left it there.

As it was, the baby arrived a week early, in perfect health, a bouncy boy of seven pounds and three ounces. He landed on Earth two hours after Neil Armstrong landed on the moon, on July 20th. He was adorable, although his light, coffee-coloured skin indicated that his father must have been one of her male Motown backing singers.

When Lottie took him out for a walk in his pram for the first time, she was startled by the hostile looks she got from her previously friendly neighbours. Illegitimate was one thing, it seemed, black something totally different. But she carried on, and the baby's beaming smile and sweet disposition won over even the most racist of the local inhabitants.

Flo went off on tour in September, heading for Australia, New Zealand then back to America. The baby, now christened Charlie, duly arrived at Wisteria House three days a week, and gurgled and giggled next to David as he wrote *This Poisoned Land*, as he had named the new story. The author was one of his godfathers. The lead singer of *The English Hooligans* was the other one. While David took his duties seriously, the christening was the first and last time the singer saw his godson. A year later he wrapped his Mini Cooper around a tree in Devon, high as a kite, and died instantly.

Henry adored the baby, and sat, tail wagging, nose on the edge of the carrycot, sniffing Flo's son, or rolling around on the floor with him if Charlie was on his blanket.

Shortly after his arrival at the house, Lottie called David and Will into the kitchen and said, 'Now look, you two, if this is going to work, you're both going to have to change him sometimes. I can't keep running downstairs every time he needs a new nappy.

Charlie was lying on a towel on the kitchen table, kicking his chubby little legs in the air, happy to be the centre of attention.

'Oh God, really?' said David.

'Yes. Now it's very simple, especially as we now have these disposable nappies. Look. You take them apart then throw it away…'

'Jesus Christ, what's that?' said Will.

'It's his little cock,' replied his grandmother.

'Little? It's huge. And… oh…that smell…'

'Yes, it's not very nice but…you wipe here, and here and… here…'

The two men had gone very pale and stepped back from the table. Lottie went on, 'Then you take this wet wipe thing and you wipe there, and there and…there. Powder there and there, then you plop him on a new nappy, and attach it there and there. All done.'

'We need to employ a nanny,' muttered David.

'No, we don't. You can do this.'

'Hmmm,' had been the only response from the two of them.

*

On the seventh of January 1970, Lottie was woken in the middle of the night by a telephone call from Austin, Texas. It was Flo's manager, informing her that her daughter had been found dead in her motel room, a syringe in her arm, having overdosed on heroin. David and Lottie flew over to bring her body home, and she was buried in St. Stephen's churchyard on what would have been her twenty-second birthday.

The church was packed with musicians and other celebrities, and Flo Nolan joined the ever-growing list of stars who had lived fast and died much too young.

Flo's two albums became rock classics, and a trust was set up to collect the royalties and ensure that Charlie Nolan would be a very wealthy young man when he turned twenty-one.

Lottie was devastated by Flo's death, but Charlie gave her a reason to keep going. She kept working at Wisteria House, and the baby became a semi-permanent member of the household. The only problem was that Henry now added Charlie's toys and dummy to the list of items to be found hidden under the rug on his bed.

Chapter Thirteen

Will started that morning by sharpening the blades of his hedge clippers. He used the wet stone to give them the sharpest blade possible, because he was going to trim back the two beech hedges that framed the herbaceous borders. As much as he loved gardening, this was one task he dreaded. It was very monotonous, back-breaking work, and would be followed by raking the cut leaves and twigs, getting them into the wheelbarrow and adding them to the new compost heap behind the vegetable garden. As each hedge was nearly sixty feet long, and he had to cut both sides and tops, it would take him two full days to complete the job.

It was a hot day in early June. David was inside writing. The baby was with Lottie in the scullery, as she was doing the ironing. Henry was laying at Will's feet, chewing a stick, waiting for him to get going.

He had been at the pub the night before, chatting to his old mates. One of them, Ned Wilson, had been complaining about how hard farming work was; he had just taken charge of his family's pig farm, as his dad had had a heart attack, a mild one, and could no longer run the place as before. He had looked at Will and said, 'It's all right for you; you just do a spot of gardening for a living.'

Will had given him the finger and tried to explain what he did all day, without the big machines they used for cutting and trimming their hedges, for example. "Do it all by hand, don't I?" They hadn't been convinced.

Satisfied the blades were sharp enough to start, he put on his thick gardening gloves to protect himself from blisters, picked up the stool and said, 'Come on, Henry.' The retriever got up and trotted after him as he walked down the side of the garage to the end of the hedge by the kitchen.

He always started there, as that side was still shaded from the morning sun. He began at the side end; the hedge was two foot wide and five foot high, so that went quickly, then he stood on the stool and clipped the top. He was five six and needed the stool to help him ensure the top of the hedge was perfectly flat. He prided himself on achieving the perfect "short, back and sides".

It was going to be a very hot day. He always cut the hedge around June, having discovered, after twelve years of trial and error, that it was the perfect time. The leaves would grow back nice and full in time for autumn, and then turn a light copper colour and stay on the branches throughout the winter. As he clipped along the side of the beech again, he thought about his garden.

He had made so many changes in those years. Like the willow trees by the pond. Back in the summer of 1966, when England won the world cup, the joy of that event had been overshadowed when he realised the trees were getting too big; they were sucking the water from the rest of the garden and damaging the pond through their invasive roots.

He finally took them both down in September that year and dug out the roots with the help of Ned and Sam, another friend, who brought his tractor through the gate by the churchyard wall at the end of orchard, using chains to drag the stumps out.

He had had long discussions with David about what to put in their place and they decided upon two maidenhair trees, *Gingko biloba*, which didn't need so much water. They had been delivered by Hillier's, a tree nursery near Winchester, two six-foot examples, one male, one female. They were doing well, and he loved their kidney shaped leaves. Gingkoes were one of the oldest trees on Earth, dating back to before the dinosaurs, but rare in Gloucestershire. Will's mental journey back in time was abruptly ended when he heard David call out, "Tea!"

He walked over to the terrace where David was now sitting at the small teak table, smoking. There were two mugs of tea and some custard cream biscuits on a plate. And an ashtray. David never flicked his cigarette ends into the garden. Will sat down, exhaled and said, 'It's a bugger doing this.'

'I know. You always say that, but it's worth it. You make it look amazing.'

The gardener smiled, took a sip of tea, then helped himself to a Benson and Hedge's Pure Gold from David's packet.

Henry sniffed around the table, trying to locate the source of the smell of biscuits.

'How's the writing?'

'Just making final corrections then it's off to Miss Bottomley for the final draft.'

'Nice. Happy with it?'

'I think so. The lawn looks great, by the way.'

Will had cut it the day before. The stretch between the two borders was perfect now, after over a decade of hard work. Straight lines ran up and down, and there wasn't a daisy in sight. It was as flat as a bowling green. David had bought him a new Atco diesel mower, with a roller on the back and a detachable bucket on the front for collecting grass clippings. It was the latest model that had a geared motor that drove the machine forward. Much easier than pushing the old heavy one.

'Ta. Yes, it does. Oh, this is nice. A hot cuppa and a smoke.'

He sat back, grinning. David nodded, stroking the dog's head, but ignoring the look of starvation on the pup's face. He was very well fed, and a custard cream was not essential to his survival.

They sat in companiable silence before they heard the phone ringing inside. David got up, blew him a kiss, and disappeared. Will took a biscuit, ate it, tossed one to Henry and went back to work. He had finished one whole side by lunchtime, then, after a sandwich, started on the other. This time he wore a bamboo hat, like rice growers used, to protect himself from the sun. Lottie had found it in a junk shop in Stroud and had given it to him as a joke, but it proved perfect. It was lightweight, wide enough to protect his shoulders if he was stripped to the waist and kept the sun off better than any other headwear. He never wore a cap or a hat normally, as he felt they made people's hair fall out, and he was very proud of his full head of hair. Ned was nearly bald already, having worn a cap almost from birth.

Will worked along the hedge, starting from the kitchen end again, but behind the herbaceous border itself. He thought about his vegetables.

The potato patch had been extended, and he now grew onions as well as cauliflowers and Brussels sprouts. He had added lines of raspberries, red currants and a large area of strawberry plants. David loved those, as did Charlie.

The fruit was beginning to ripen and Will had laid down straw under each plant to protect the strawberries and keep them clean. One thing they no longer had any problems with were rabbits. When he had started in 1958, he had had a running battle with the rodents. Despite Flush's best efforts, chasing off any that he caught eating on the lawn, they slipped into the walled garden and ate his lettuces and carrots, nibbled the tender shoots coming up in the borders in spring, and left their tell-tale droppings everywhere. When myxomatosis, the rabbit disease, finally reached the West Country
in 1959 and 1960, it had decimated the population around Stroud and its villages. It was a vile illness, that produced the classic, white-eyed effect dying rabbits had. While Will was glad to not have to fight them off his vegetables, they had often came across dying rabbits by the roadside, and it was very sad and distressing to see them suffering. David took to carrying a big stick for a while to save them from further pain.

Other mammalian visitors were more welcome, like the hedgehogs and the occasional badger. They had a family in the beech wood, that had dug a set into a bank near the farthest end. The fox wasn't so welcome, especially after it took a chicken. They had mice too, of course, everyone did. But they didn't want to have a cat, so they left the mice to the neighbours' moggies to deal with.

As he kept on clipping, he smiled to himself. He loved this garden, and he loved his work. Despite this particular job. He wasn't sure when it had happened but gardening had become more than a job, it was now a hobby and a passion.

David had said they were a team, making a home, and he had total freedom to try new plants, redesign the layout and make it whatever he wanted it to be. He grew wonderful flowers and exceptionally good vegetables, which he was very proud of. However, he never exhibited them.

Fletcher's Cross, like most English villages, always held a summer fete on the village green, in aid of the church's constant renovation fund. There would be games like ten-pin bowling, archery and a coconut shy. Stalls would go up selling home-made jams and cakes, a tea tent was available for refreshments, and a big red and white striped marquee for the competitions. There was a dog show, a small pets show and, of course, a flower and vegetable show.

When David asked why he didn't want to take part, he'd replied, 'Ain't one for showin' off, Davey.' Actually, his main reason was one of privacy. He didn't want to get his picture in the papers, even for best onion or carrots, in case people started to wonder about him and his lover.

Will finished the first hedge at four, then raked up all the bits, making several trips to the compost heap. He added the cut twigs and leaves to the growing piles of grass cuttings, dead rose heads and other garden waste which would eventually produce a rich compost to be dug into the borders and vegetable patch, along with manure, in the autumn.

He was exhausted now, with aching shoulders and was sweating like a pig. And he would have to repeat the whole process the following day to clip the second hedge. Even though he was tired, he took a moment to fill up a watering can and went into the greenhouse for his final task of the day, watering his tomato plants. They were growing well and he would have a fine crop that year, as usual.

He went in through the back door and, as he was taking off his boots, he smelt the wonderful aroma of frying sausages. David was making him sausages and mash, his favourite meal. As he went into the kitchen, his lover smiled and said, 'Thought you deserved a treat after all that work.'

'Lovely. I'll just take a quick shower then I'll be down.'

He kissed David and headed upstairs, his working day over. As he stood under the hot water, letting it ease away the pain from his muscles and wash away the sweat from his body, he laughed. His mates thought he had an easy life. Just let them cut that bloody hedge, he thought. That would show 'em.

*

One afternoon the following week, they were sitting on the bench by the pond, having a cigarette, at the end of another working day. It was very peaceful, and warm, and they were both relaxing after their labours.

Even though they had lived together for over twelve years and would soon be going in to have dinner, they often met up like this for a smoke and a chat, but now as friends and lovers, not employer and employee.

David was looking across the potato patch towards the orchard. There was a fence made of hazel sticks, a woven wattle affair, stretching from the hawthorn hedge at the rear of the garden right across to their neighbours' back garden, with its cedarwood panelled fencing.

This hazel barrier kept the sheep from roaming around the rest of the garden. For the last three years, Will had borrowed a pair of sheep from a neighbouring farm to keep the grass down under the apple trees. It was easier than cutting it all the time, and they never caused any problems. They arrived at the end of March and would be collected, for whatever future destiny had in store for them, at the end of September.

David had thought it had been a brilliant idea and liked to lean over the wattle barrier and watch them graze under the trees.

He stopped looking at the fence and turned his attention to the magnolia and the gingkoes. They were three wonderful trees but caused Will a few problems. In the spring, the *Grandiflora* bloomed, huge pink flowers that, after a while, turned cream and dropped off into the pond. In the autumn, the tree's thick oval green leaves turned many shades of brown and followed the petals into the water. The gingkoes did the same, covering the surface with kidney-shaped cream leaves.

Will now used wire netting twice a year to cover the water and collect this detritus. Behind the magnolia was a strip of uncultivated land. This ran from the orchard, behind the pond and the heather garden, right up to the rhododendron walk. It was a strip about thirty yards wide, running along the hawthorn hedge. That divided their land from Farmer Robson's field.

The grass was neatly cut just behind the magnolia, but then it grew wild, like a meadow. Poppies, lupins and other meadow plants were growing amongst the long grass.

'That's working well, the meadow bit.'

'Yep, it is. I've kept the dandelions out; that took a bit of work as they seed like crazy, but now it's pretty good. Got some nettles and thistles in there too, like Peter Scott told us to, when we first went to Slimbridge.'

'I remember. And the bird boxes have done well. Loads of tits and flycatchers nested this year.'

Will nodded. 'He's clever bloke, that Peter is.'

'Will you keep it like this?'

'I think so. Apart from the fact it cuts down my lawnmowing, it really looks beautiful. Well, I think it does.'

'I do too. Natural. Like a proper meadow. You're bloody good at this, you know that don't you?'

Will smiled. He liked it when David noticed his efforts, and his expertise.

'Ta. Yeah, maybe I am.'

'Don't be so modest, love. This place is a real tribute to your skill.'

Will gave him a mock withering look then said, 'You'll be wanting sex tonight then, will you?'

David laughed. 'I'm not chatting you up for a shag. But yes, I'd love to take you to bed again. Right now, if you're in the mood.'

Will felt himself growing hard at the thought. He bit his bottom lip and said, 'I'm really sweaty. Want me to 'ave a shower first?'

David stubbed out his cigarette in the stone ashtray they had placed by the seat and got up, holding out a hand.

'No need. I'll take you as I find you. And I'll take you nice and hard if you like.'

Will got up, took the proffered hand and nodded. 'I like. Very much…'

Chapter Fourteen

On the tenth of July, David received the completed, typed draft of the new book from Miss Bottomley and started giving it a final read-through in the garden. The weather was glorious, and Charlie, nearly a year old, and wearing a very large bucket hat made out of blue canvas, was laying on his belly on a blanket next him, playing tug of war with Henry, the baby's teddy bear being a substitute for rope.

Will was in the vegetable garden, tying up his runner beans, and Lottie had just finished the bedroom and bathroom and decided to make tea and have their elevenses outside. Carrying a tray with three mugs of PG tips, a plastic mug of milk and a plate of McVities chocolate digestives, she made her way across the lawn to the table next to the author.

'Oh, move those papers, love.'

He shifted the manuscript's box onto the lawn and called out, 'Will! Tea.'

Lottie picked up her grandson and held him close as she helped him drink his milk.

'Ooh, he's getting heavy.'

'He's a big lad. You look after him very well.'

'Well, he's all I've got left.'

Will, coming across the grass, said, 'You've got us too, Lottie.'

'Oh, you know what I mean.'

'I do. Oh, digestives…'

He took two and sat down next to the dog. Henry rolled over, yapped, waiting to be scratched on the belly. Will duly performed this happy duty, and Charlie, back on the blanket, waved and gurgled at his furry friend, who started tugging on the teddy bear, renewing their battle.

'How's the book?' asked Will.

'She's done a great job as always. Do you know, I've never actually met Miss Bottomley? She answered an advertisement I put in the Oxford Times back when I was at university, and started typing for me then, but we have never met.'

'Perhaps we should invite her over.'

'The problem is, she is in a wheelchair and I am not sure how well she would cope with the house. Or this garden.'

'Where does she live?'

'Swindon.'

'Well, let's find a restaurant near her that can accommodate a wheelchair and we can take her out to lunch. Or you can if you don't want me along.'

'Of course, you can come. That's a very good idea. I'll ring round a few places in Yellow Pages this afternoon then give her a bell. I wonder what she is like?'

Lottie asked, 'You know nothing about her? You've worked together for over twenty years!'

'I know. Madness, really. I've written to her and we've spoken on the phone many times but never met in person. Will, you'd better rescue Charlie.'

The dog, now almost fully grown, had dragged the baby off the rug and was pulling him towards the pond. 'We're going to have to cover that with wire netting once he starts walking…'

'Soon. He can crawl faster than any of us can run now,' muttered Lottie.

David found a hotel that was adapted for visitors in wheelchairs just outside Swindon, in a village called Stanton Fitzwarren, the Stanton House Hotel. It had a fine restaurant with three rosettes from the AA. He called Miss Bottomley, who was surprised and delighted at the idea, and they booked for lunch the following Wednesday. She could, she said, get a taxi there, as she had an arrangement with a local taxi service that knew her well and dealt with her chair in a very professional manner.

Before he finished the call, David said, 'I hope you don't mind, Miss Bottomley, but I'll be bringing my friend Will with me. He's dying to meet you.'

There was a long pause before the cultured voice, with a trace of humour, asked,

106

'Is he a very close friend, Mr Manners?' with a heavy emphasis on the word "close".

David coughed and said, 'Um, he is rather.'

'In that case, may I bring my close friend too? Her name is Mary-Louise.'

'Miss Bottomley, of course you can.'

'She'd be thrilled to meet my most famous client.'

'It seems we have more in common than just my books.'

'Indeed, we do, Mr Manners. See you next Wednesday.'

He went into the kitchen where Will was soaking Henry's meal.

'Turns out she's a lesbian. She's bringing her friend with her.'

'Well, this will be fun, won't it?'

'I rather think it will, as she sounds very amusing.'

It was. They arrived early and were standing waiting when a Ford Zodiac taxi parked next to them. A tall, grey-haired woman got out of one side, as the taxi driver fetched the folding wheelchair from the boot. He helped a tiny, bright-eyed woman from the other rear door and Will muttered, 'Gosh. She looks as if she should be investigating murders in St Mary Mead.'

Miss Bottomley did indeed look like Miss Marple, Agatha Christie's famous female detective. She waved to them as she settled into the chair then thanked the driver, who said he would collect them at two-thirty. As they walked up to her, she said, 'This is my friend, Mary-Louise Browning. And I am Diana Bottomley.'

'And this is Will Forman. I'm David Manners, Miss Browning.'

'Mary-Louise, please, Mr Manners, although I can't help but think of you as Connor Lord.'

He laughed and then wheeled his typist into the hotel, where they had drinks in the lounge before going into the restaurant. They discovered that Miss Browning was the head matron at Swindon General Hospital, and a huge gardening enthusiast. It didn't take long before she and Will bonded, discussing the best way to remove earwigs from dahlia flowers, and the best time to prune apple trees.

Diana and David focused on his books. He said, 'I am amazed that you manage to make sense of all my notes and scribbles when I send you the handwritten chapters. They seem very messy to me, but you manage every time.'

'After twelve books, I know what you mean to say, even if it is just a jotting in a margin. You have a very distinctive style and I can predict what you mean now very easily.'

'Really?'

'Oh, not the story lines… they are always a complete surprise. But once you start a sentence, I know where you are going and what you want to say.'

'That's wonderful. Although *Storm* must have tested you to the limits.'

At four hundred and sixty pages, and one hundred and eighty thousand words, it had been twice the length of his previous novels.

'Oh yes. Do you remember when I called you about that?'

'Like it was yesterday. You asked if it would ever end, and I said that yes, the world as we knew it might come to an end…'

'And I said, "No, Mr. Manners, I meant this book!".'

They chuckled at the memory and he looked over at Mary-Louise and asked, 'Um, how long have you two been together?'

'Since 1938. We met when I was in hospital suffering from a bout of pneumonia: very dangerous if one is paralysed. I was already in the chair by then. Mary-Louise was training to be a nurse. We were both eighteen and fell for each other on sight.'

'How very romantic.'

'And you and Will?'

'Just over twelve years now. He came to work for me as my gardener. Still does.'

'Has the change in the law helped things?'

'The 1967 Act? In one way, yes, but not really. We can do what we like in the privacy of our home, but we still need to book two hotel rooms, and can't hold hands in public. There's a long way still to go.'

Miss Bottomley had referred to the Sexual Offences Act 1967, which had finally legalised homosexual acts, as long as they took place behind locked doors and closed curtains in the privacy of one's home, and if both men were over the age of twenty-one. Any public signs of affection were still met with jeers and derision, and the police had started using agent provocateurs to catch any man trying to have sex in public toilets and parks. The number of cases before the courts had risen sharply in the three years since the new law had come in.

'I know. Why do the police target homosexual men so much?'

'There is still a great deal of prejudice and fear. And disgust, to be brutally honest. But I agree, considering there are so many other crimes to be investigated, they do spend far too much time focusing on lonely men trying to find a bit of fun. But I wish there weren't so many hanging around toilets. It doesn't help matters.'

'But where else can they meet?'

'That's true. There are only a very few pubs, and they are all in the big cities. There are none around here. It's not like America.'

The Stonewall riots had taken place the previous year; New York and San Francisco had many bars for homosexuals, even if they were mainly owned and run by the mafia. The idea of "rights" for gay people were moving forward, albeit very slowly.

Diana nodded and said, 'It's much easier for us. No-one thinks twice about two old spinsters living together. Many women do, you know.'

'I know. In long, loving faithful relationships too.'

'Are you faithful, David?'

'Utterly. Never even thought about cheating on Will. He's adorable.'

She laughed, 'He is.'

'My ears are burning. What are you two talking about?'

'You, love. Oh, that looks good.'

The waitress had placed a fine dover sole in front of him, with fresh tiny brown shrimps sprinkled over the top, along with a little jug of hot butter. He asked, 'Have you eaten here before, Diana?'

'Once. On my birthday two years ago. I was delighted when you suggested it. The food is excellent.'

'But bloody expensive,' said Mary-Louise.

Will turned to David and said, 'We've been discussing a visit to Wisteria House. Mary-Louise thinks Diana can manage with the chair, and the downstairs loo is big enough for her if it is needed.'

Diana blushed. 'Well, really. That's a fine topic to have over a nice breast of duck...'

They started on their main courses, and David offered a toast. 'To the best typist in England.'

'Oh, don't be silly, dear boy. Eat your fish.'

Her friend asked, 'David. Your last book, *The Endless Storm*. Do you think things could really get that bad one day?'

He put down his knife and fork and nodded. 'In some ways, maybe even worse.

In her book, Rachel Carson talked about the death of birds caused by DDT and touched on the death of bees and other insects because of pesticides. I only briefly covered it, but one professor I spoke to said that, if the bees die, we all die. Nothing to fertilize the crops, you see. Unless we are very careful, we could be creating a future just as bleak as mine. And that's before we start on the changes in the weather.'

'That is so sad. Diana mentioned you had contacted some members of parliament as you were writing it, to see how many understood or were considering the issue.'

'I did, both Conservative and Labour MPs. About twenty in all. Not one of them had read *Silent Spring*. None of them thought the environment was a big issue, and a couple wrote back saying they were more worried about what was going to happen next week than what might happen in thirty- or forty-years' time. But if we don't start acting soon, that time will rush towards us and it might be too late.'

'I thought your description of the destruction of London was truly awful. I found that very disturbing and very real.'

He smiled sadly, 'I'm so sorry. But that bit was based on some very detailed research. And, mentioning how long it takes for politicians to act, they have been discussing building some sort of barrier across the Thames ever since the North Sea Flood in 1953. Then thirty thousand people had to be evacuated from their homes and over three hundred were drowned along the east coast. Parliament itself was partially flooded and there was a real danger of people drowning in the underground.

Here we are, nearly twenty years later, and still nothing has been done. They are designing something now, but it won't even start construction for another couple of years at least.'

'It is very depressing, but it made for a wonderful read.'

He laughed. 'Thank you, ma'am.'

'As did the newest one. I hope you don't mind, but I always read the finished manuscript before Diana sends it off to you.'

He smiled, 'Not at all. It's a bit more local than the last one.'

'Oh, I think it benefits from that.'

Will looked confused. He never read David's books until they were properly published, and even then, he struggled sometimes, as he was slightly dyslexic.

He asked, 'What's it about?'

The two ladies looked very surprised. 'You don't know?'

'No. I leave David alone when he writes so it will always be a surprise when I see the finished book. He always shocks me too. There he is, a nice-looking gentle Englishman writing at his desk, and he produces these terrifying books about murders and secret agents and now the end of the world. I wonder who I'm living with at times.'

'He does have a very good imagination.'

Will grinned, 'He does, doesn't he? So, go on, this new one?'

'It won't spoil it for you?'

Will shook his head, 'Go on.'

David said, 'Well, it is set in Manchester, which is why I went up there a few times as you know. Someone in a chemical plant accidentally releases some very nasty new toxic chemical compounds into the water supply, and the book is all about what happens afterwards.'

'Oh. After the end of the world that sounds a bit… dull.'

He said it with a cheeky smile on his face and David rolled his eyes. But Mary-Louise came to the author's defence.

'Not at all, Will. It is so good. The way things develop; the hospitals overflowing with patients, the sickness that makes people go mad with rage, the way the company tries to avoid blame. The local politicians, trying to make hay from the disaster, then back-tracking like mad when they realise some of their decisions led to the mess. The government acting far too slowly. It is very, very good and deeply disturbing too.'

'The sickness that makes people go mad. Is it a zombie story then?'

They had been to see *The Night of the Living Dead*, directed by George A. Romero, when it had come out a couple of years before, and Will had been unable to sleep for several night afterwards. As much as he loved science fiction, he wasn't a zombie fan.

Miss Browning said,

'No, it's far cleverer than that. Oh, the scene at the football match, David. That was wonderfully nasty.'

He laughed at her enthusiasm for the macabre then looked at Will and said, 'I wrote about a football match between the two Manchester sides, where the supporters, affected by the tainted water, attack each other in the stands. Glad you liked it, Mary-Louise.'

'So realistic, even it was a bit wild.'

The waitress returned to clear the plates and they ordered coffee to go with the sweet trolley. Both ladies ordered the trifle, and he and Will had crème caramels. They were almost as good as David's. After lunch, they sat together at a table outside the main entrance as they waited for the taxi to arrive.

'I'm so glad we have finally met, David. It really has been a pleasure.'

'Totally agree. Crazy that we left it so long. And I am looking forward to you both coming over in August.'

They had agreed to come for a Saturday lunch early the next month.

Diana asked, 'Can we bring our dog?'

'What is it?'

'She's a Border terrier. Only two years old. We have always had those. She's our fourth.'

'Please do. We have always had dogs too. We love them. We have a golden retriever at the moment. He's a bit boisterous but very sweet.'

'Oh, Mooky loves other dogs.'

David raised one eyebrow and replied, 'Interesting name.'

Diana laughed, 'Silly, made-up one, I know, but she looked like a Mooky when we fetched her from the breeder.'

'Ours is a Henry,' said Will.

The conversation had to end there, as the taxi arrived. They helped the two ladies into it and waved them off.

On the drive home, Will asked, 'Davey? What does a mooky look like?'

He laughed. 'God knows, but we're going to find out in a couple of weeks...'

Chapter Fifteen

They arrived at Wisteria House, not in a taxi this time, but in Miss Browning's big metallic blue Humber Spectre. Will helped Mary-Louise get the folding wheelchair from the boot, as David got Diana out of the front passenger seat. They then wheeled her into the house. Mooky, a sweet Border terrier, was pounced on by Henry, and the two dogs ran out through the living room into the garden, rushing around, barking happily.

They settled on the terrace for drinks before lunch.

It was a swelteringly hot August day, welcome after several days of rain. The garden looked lush, the herbaceous beds glorious, and the apple trees heaving with ripening fruit. Mary-Louise and Will went off for an inspection, and David and Diana sat chatting. The dogs, having taken a drink from the pond, finally relaxed and lay down panting next to their owners.

'I've sent the book off to the publishers. My editor likes it very much, but he's a bit worried about drinking tap water after reading it. He was concerned I might put people off it for life.'

'Oh, but that's why it was so effective, David. Something so every day, that we take for granted, becoming the vector for terrible health problems. You mustn't change that.'

'I won't,' he said, laughing. 'I must ask you for your opinion from now on. I should have done so long before.'

'That would be nice, but I would never presume to influence your work. It is so original. You always surprise me.'

'Good. That's the general idea.'

'How many copies of *The Endless Storm* have you sold so far?'

'Over twelve million world-wide. A million in hardback, the rest in paperback. Still selling well at airports. It's become a classic holiday book.'

Package holidays had taken off during the sixties, and many travellers bought a book at the airport to read on the beach. *Storm* was now one of those big fat paperbacks that everyone seemed to have on their bookshelves, like *War and Peace* by Tolstoy, and Tolkien's *Lord of the Rings*.

'Twelve million. Gosh, that's wonderful. And all your books?'

'Over fifty million now. Not up to Dame Agatha's level quite yet but I'm getting there.'

Agatha Christie was still Britain's best-selling author, with eighty million books sold, and was the one all writers hoped to emulate.

Diana gasped, 'So many? That's wonderful.'

'And you have been a part of that success. I couldn't have done it without you.'

'Well, I've been well rewarded. What will you write about next?'

'Pharmaceuticals, I think. I was reading about the Thalidomide case again recently. You know the Sunday Times is trying to get more compensation for the victims? I think I could get a good story out of such a case. But with an American angle this time. The lobbyists in Washington, for example. I had a reader contact me about how much money drug companies give to American politicians to stop them looking too closely at their activities. And to stop them developing a National Health Service in the States. The insurance companies want to keep selling medical insurance to companies and families. It's big business over there.'

The drug Thalidomide had been produced in the 1950s as a sedative but was then found to help women during pregnancy with morning sickness. But some mothers had given birth to babies with terrible defects, like short limbs, or truncated bodies. Their struggle to get money in compensation had taken years.

'Gosh. Well, they have now changed the rules, haven't they? Drugs for people have to be tested on people, not just animals, before they can be sold.'

'True. But there are loopholes. I want to expose those.'

'Sounds very exciting. Can't wait to read it.'

'Year after next at the earliest. I am going to have to do an awful lot of research to get it right. Well, Mary-Louise, what do you think of Will's Garden?'

The two gardeners had returned. Mary-Louise sat down and he topped up her lemonade from the jug, clinking with ice cubes. The rest of them were drinking Pimm's. Matron was driving.

'I think it is wonderful. Such a mixture of wild and controlled, flowers and herbs, vegetables and shrubs. And the greenhouse is amazing. I've never seen avocados growing in England before.' David smiled. 'It's taken a bit of time…'

Will gave him a dirty look. His struggle to produce the fruit in the greenhouse was a long-standing joke between them, but he had finally managed to get one of the plants to produce one single avocado.

He also grew lemons, which did very well, and the roof of the greenhouse was now covered by the vine, with bunches of grapes forming all along it.

The lunch proved a big success, and the four became firm friends. With *This Poisoned Land* finished, David had to find a new plot. He considered the problems in Northern Ireland between the protestants and the Catholics; he toyed with the idea of writing about the Yorkshire Ripper murders, still under investigation, and even thought about the Vietnam War, which the current US President, Richard Nixon, was trying to get out of. As usual, he would go walking, taking Henry now, but the dog was so young and playful, he found himself watching the pup more than thinking about a story.

Finally, one day in November, he was in Boots the Chemist in Stroud, waiting to buy some aspirins when he overheard an old lady talking to the man behind the counter. He had just got her new prescription ready and, as he gave it to her, she said, "Oh, good, now I can keep on living. This stuff keeps me alive, it does." In a flash of inspiration, he had his new idea. He got the title first, *The Perfect Cure*, but the plot followed very quickly afterwards. When he got home, he started on his research and was soon buried in articles and materials about the pharmaceutical industry.

That December, they invited Mary-Louise and Diana back to Fletcher's Cross to join them on Christmas Day, along with Lottie and Charlie. Their visitors from Swindon came by taxi, so they could both drink, and they brought a magnificent box of Milk Tray chocolates with them as a gift. David and Will gave them a very fine bottle of brandy, and Lottie got a Robert's radio, a very expensive model at the time, as she and Charlie liked listening to music when she knitted at home. She gave them both new pairs of socks.

Charlie got a huge toy dog on wheels, which sent Henry and Mooky into hysterics, barking and pouncing around it as Charlie pulled "Doggie" all around the living room.

It was a social and peaceful end to the year, and the start of another decade in Fletcher's Cross.

Chapter Sixteen

This Poisoned Land was another huge success for David and for his publisher, Norman Hilton. His company was doing well, and he bought up a couple of other small publishers and their back list of authors. David's agent, Murrey Holt, was joined by his son Morris, who had just graduated with a degree in marketing from Warwick University, and they took on several new clients, as Murrey's reputation, thanks to Connor Lord's success, had made him the hottest agent in town.

At Wisteria House, Charlie Nolan started walking, then running. They covered the pond with wire netting, then took him to the public swimming pool in Stroud to teach him to swim so they could remove it again. When the time came, Will taught him how to ride a bicycle. They sometimes took Lottie and her grandson to the seaside, a strange little family group of an older woman, two men and a black child. Charlie was sometimes subjected to nasty racist comments, shouted out across streets or on the beach, but David told him ignore them.

'They're just jealous you look so cool.'

Charlie was fascinated by the piano. As a toddler he would stand, hands on the piano seat, and watch as David played. He loved Nina Simone's *My baby just cares for me* most of all, and the author would play the intro many times as the boy chuckled and waggled his bum to the music. He loved the "De-dum-dum-dum-dum-dum-dum-dum-dum, De-dum-dum-dum-dum-dum-dum-dum-dum" rhythm. Lottie and Will would often stand in the doorway, watching them. Once Lottie turned to Will and said,

'He's a natural father, ain't he?'

'Looks like it.'

David and Will used to go up to London at least once a month, leaving Henry with Lottie. He had grown into a very big, happy dog, and Charlie adored staying at the house with him. In town, they started to visit one or two of the pubs that other gay men went to, albeit very discreetly. The Boltons, on Earls Court Road, was their favourite.

They would chat with other drinkers, but never tried to bed any of them.

Nor allow themselves to be bedded either. They would sometimes meet up with Paul Ford, David's editor at Hilton Publishing. He and his boyfriend would take them around, and they got to know several other gay men, so were not total strangers when they went to the bars. Paul and his partner had an open relationship and often picked up a lad to take home with them.

The customers were a mixed bunch. Some were very flamboyant and loud, others more discreet. Many arrived looking very nervous, on their first visit to a gay bar, only to return later more confident and relaxed. Men of all ages went to the Boltons, and there were several couples where a much older man stood with his arm across the shoulder of a younger lad. Sugar daddies and their protégées.

David and Will went to the theatre as well, seeing some of the best shows of the decade, including musicals like *Jesus Christ Superstar* and *The Rocky Horror Show*, and dramas like *Sleuth*, with Laurence Olivier.

Back home, they still watched TV, especially *Top of the Pops*, to keep up to date with new music. David loved Cat Stevens, and Will fancied Marc Bolan from T-Rex. They bought a lot of albums. When Simon and Garfunkel released *Bridge over troubled water*, they played it endlessly, and David would sit at the piano and play several of the tunes as well.

They also discovered a very odd singer called David Bowie, who produced one of the strangest albums they had heard in December of 1971, called *Hunky Dory*. David loved the track, *Life on Mars*, and would often play it on his piano, very softly and slowly, as he took breaks from his next book.

They even went and saw him in concert; they looked a bit out of place with the rest of the crowd, in eye make-up, spangly, sparkling tops and tight black jeans, but Bowie was totally amazing, so they ignored the disapproving looks from Bowie's fans, and just enjoyed the music and his performance.

Will was developing the garden during this period. They had a very dry summer in 1972, with a ban on using hose pipes. He found and ordered water butts for all the down pipes, on the house and the garage gutters, to collect any rainwater in the future. He had extended the vegetables again, and was now growing shallots and black currants, had developed the asparagus despite never eating any himself, and added new pear and cherry trees to the orchard. He also learnt to drive so he wasn't dependent on David all the time. When he passed his test, they bought a second car, a Morris Traveller, the one with the wooden strips on the side and double doors at the back. He'd asked for one so he could go and collect plants and bags of soil from the garden centre in Stroud. Henry loved this car and they used it often when taking him off for the day. Charlie liked it too and would sit in the back on Henry's blanket, cuddling the big dog, whenever they took him out to give Lottie a break from her very energetic grandson. It was before child car seats became law and it was lucky that they were never in an accident, as both the dog and the boy would have been thrown all over the place had they hit something.

They became Charlie's surrogate dads over the next few years, teaching him how to weed the garden, play chess, play the piano, and read and write. David had finally become a father, despite saying he didn't want the role after Eve's death.

*

The day after Diana Bottomley and Mary-Louise had visited them for the first time, David took Henry for a walk through the heather garden to the beech wood. He loved walking here, as it was so peaceful. The retriever found a big stick, and trotted ahead of him, proudly carry this trophy in his jaws.

It was another hot day and the ground was very dry underfoot. Will had cut down a rotten beech two years before, which he had cut into logs, but he had left one big piece of the trunk laying on the woodland floor, as a seat. The author sat and breathed in the fresh country air, looking up at the canopy of deep green leaves that covered the glade that had been opened up by the dead tree's removal. Henry lay at his feet, chewing on the stick.

Dappled sunlight filtered through, creating a magical atmosphere.

Clearing a small area of dirt with his shoe, so he could stub it out later, David lit a cigarette. He watched the smoke drift up through the sunbeams. A wood pigeon cooed up in the branches overhead, and a pheasant clacked from the field behind the hawthorn hedge. He was so damned lucky, he thought. He owned a wonderful house and had a much-loved partner, who loved him back. He was successful, financially secure, and he owned a woodland. He owned a wood! Not many people could say that.

He had loved coming here from the start, first with Flush, and now with Henry. The only time he had had problems was when Flush had spotted two roe deer, feeding off the beech nuts one autumn, and bolted after them as they sprang away. It had taken him over two hours to find his Labrador, and only then by taking the car and driving across half the county. He had finally located the damn dog, tail between his legs in contrition, standing in a lane near one of the farms. He had been exhausted by the chase, covered in mud, and shivering with cold.

That day, as he sat there in peace, David started thinking back over the last few years, since he had arrived in Fletcher's Cross. He had been so happy here. Meeting Will had transformed his existence, and the warmth and good humour of the locals had only added to his pleasure of being part of this small community.

He had, he thought, only two big regrets in his life. The first was that it had taken Eve's death for to him to move to Fletcher's Cross in the first place and find true happiness. He had genuinely liked her; they had become very close friends, and they might have made a go of things with a child had she lived. But he knew that he would never have ended up in a relationship like the one he had with Will if she had still been around. Maybe he would have found happiness with another man in a different kind of relationship if she had lived, but the joy he felt everyday with Will was partly due to Eve's absence.

His second major regret was that he had not done more personally in the fight to make homosexuality legal.

The process had begun before he moved to the village, with the Wolfenden Committee report, published in 1957. Set up to examine the laws concerning homosexuality in men (there were no laws on the statute books about women, going back to Queen Victoria's time, when she had simply said that there was no such thing as a lesbian).

The report argued that what went on in private, or what some people might consider sinful, were not necessarily criminal offences. It summarised its findings thus, *"unless a deliberate attempt be made by society through the agency of the law to equate the sphere of crime with that of sin, there must remain a realm of private that is, in brief, not the law's business."*

The very Conservative Prime Minister, Harold Macmillan, scared of a public backlash, did not act on the recommendations when the report was published. Homosexuals had to wait until 1965, when three politicians sponsored a private member's bill, the Sexual Offences Bill, which brought attention back to the original Wolfenden report. They were Lord Arran, Leo Abse and Humphry Berkeley. A majority of MPs were sympathetic, and Berkeley's bill passed its second reading in February 1966.

However, the General Election prevented it going forward then, and Berkeley lost his seat; his constituents didn't like his support for homosexuality. Labour increased their majority and Abse, one of their MPs, re-introduced the bill.

Harold Wilson's government was being pressured to bring in several social reforms, including legalizing abortion, making divorce easier, abolishing theatre censorship, and removing the death penalty. The Sexual Offences Act, that decriminalized homosexuality in private between adult men over twenty-one, finally became law in '67.

David had supported the legalization on abortions, even writing to the Times about the dreadful ways women who wanted, or needed, such a procedure, had to go about it. It was, in his words, a class problem. Rich women, albeit illegally, could get an abortion smoothly in expensive clinics, but poorer women had to rely on backstreet operations, often in very unsanitary conditions, leading to many deaths.

But he had held back from writing in support of changing the laws on homosexuality. This was partly because of the vehemence of those who opposed any such a change in the law, but mainly because of his position as his family's breadwinner. If the public had stopped buying his books, he, Will, Lottie and Charlie would all have been in trouble.

When Berkeley lost his parliamentary seat, it reinforced this fear, so David kept quiet. But he was ashamed that he had.

Sitting in the beech wood that day, he made a personal vow. There was still a long way to go before homosexuals got equal rights and became fully accepted. He would do more, much more, to try and make that happen from now on.

1980

Chapter Seventeen

'Come on, Davey, it's about to start.'

'I'm here, I'm here. Don't panic. Is the video recording?'

'Yes. Sit down, for God's sake.'

David handed Will a glass of wine and joined him on the sofa. The living room curtains were closed, and a fire burned in the hearth, giving them much needed warmth after the wet and miserable Sunday they had had. It was ten o clock at night. They cuddled up in front of the TV as The Thames Television ident came up, then the familiar intro to *The South Bank Show* started to the strains of Paganini's "24th Caprice", taken from Andrew Lloyd Webber's Variations album, playing over the now famous animated detail from Michelangelo's Sistine Chapel painting of the Hand of God giving life to Adam. The cellist playing the theme was the composer's brother, Julian. The programme was that rare thing; an art's show that people actually watched, presented by the extremely suave Melvyn Bragg.

Bragg appeared in a studio setting, and his nasally voice announced, 'Tonight, *The South Bank Show* is devoted to Britain's most successful living author, Connor Lord. With book sales of over one hundred million copies world-wide, he has been writing for thirty years but has only given two brief interviews before, back in the sixties. I spoke to him earlier this summer in the garden of his home in the Cotswolds.

The image changed to a relaxed David, in a deep blue cotton shirt and jeans, sitting at their teak table in the garden. He looked younger than his forty-nine years.

Bragg was sitting opposite him. Behind the writer, the herbaceous border was a riot of colour, a glorious reminder on that wet day of the summer that has passed.

The presenter went on, 'Welcome to *The South Bank Show*, Connor. Why haven't you given more interviews?'

Will whispered, 'Oh, you look great.'

'Shsssh.'

On screen the author smiled and replied, 'Well, I live a very quiet and some might think rather boring life. I prefer my books to do my talking for me.'

'You started writing spy fiction and were first published when you were a nineteen-year-old student at Oxford.'

'Back in 1949, yes.'

'At the same time as Ian Fleming started writing his James Bond novels. Did you feel you were competing with Fleming?'

'No, no,' David laughed, 'there was room in the market for two spies, and Ian's books were rather different to mine. The Wright stories showed the gritty, dirty side of the business, his were rather more exotic, but I loved his books.'

'You wrote to each other, I believe.'

'Yes, often. I first wrote to him in Jamaica after *Goldfinger* was published to congratulate him. I thought it was, and still is, his best book. The scene where Oddjob painted the girl who has betrayed Goldfinger in liquid gold... I thought that was so extraordinary I wrote to him and said I wished I had written it. He wrote back and said he wished he had written all of mine. That was very generous of him. We carried on corresponding until just before his death in 1964. He died far too young. Very sad.'

'Now Bond has been on the big screen since the mid-sixties. Do you like Bond films?'

'The first three were pretty true to the books, but they have evolved since then. Still a lot of fun and great entertainment, but no longer much to do with the original novels. I love seeing them, tongue firmly in cheek, but they aren't Ian's books anymore.'

'Is that why your Wright books were never made into movies?'

'Partly, yes. Back in the day, an American producer did approach my publishers wanting to put Wright on the big screen, but he wanted to make him American, and I refused. Then the BBC agreed to adapt them and did a wonderful job.

Very true to the books. As an author, you want your stories to be told the way you wrote them. Well, I do.'

'Is that why *The Endless Storm* hasn't been filmed yet?'

David smiled again. 'No, that's because the technology doesn't exist yet to film it to its full potential. We went and saw "*Star Wars - The empire strikes back*" earlier this year, and the special effects are amazing, so we are moving in the right direction. I think, if *Endless Storm* is ever to be filmed, it needs to be done well to give the same impact as the text. If it had been made as a shaky science fiction B-movie, it would have been very sad, and ruined it. It's a bit like *Lord of the Rings*. The technology isn't there yet to do justice to Tolkien's extraordinary novel. One day it will be, and hopefully a true fan will make a wonderful version, although he or she might need a couple of films to cover the whole story.'

'You like Tolkien?'

'Oh yes. He created not just a wonderful saga, but a whole new language to boot. Incredible imagination. But then I like several authors, and I am always surprised and impressed by original ideas.'

'Such as?'

'Well, H.G. Wells' *Time Machine* is a classic example. That led to many time travelling stories. He was the father of time travel. We wouldn't have had *Doctor Who* without Wells. And talking of the good Doctor, the idea of regeneration was brilliant. It has allowed the programme to renew itself time and time again. Very clever.'

'Any other writers you like?'

'When it comes to spy novels, I think John le Carré is superb. *Tinker, tailor, soldier, spy* was a stunning book, and I thought the BBC did an excellent job with their adaption last year. Smiley is a wonderfully written character. And I love Conan Doyle, of course. Sherlock Holmes is an old favourite. I admire John Irving. *The Hotel New Hampshire* was very beautiful. I was incredibly moved when he wrote about Sorrow, the stuffed dog, floating in the water after the plane crash. And I've just finish reading Maya Angelou's *I know why the caged birds sing*. A very powerful novel. I loved that.'

Bragg nodded; he had read it too and been deeply touched by the black poetess's story of her early life in poverty.

'*Storm* is your most successful book to date, and still sells hundreds of thousands of copies each year. Why is that do you think?'

'Well, it is still relevant and becoming more so year by year. We have done very little to stop the effects mankind has had on our environment, and anything we have done has been done very slowly. Take the Thames barrier, for instance. We knew, after the Great North Flood of 1953, that London would need protection in the future. Today, we have only just started construction of the barrier and it won't be finished for another three or four years. Politicians think short term, and problems many years in the future are side-lined for more current issues, but the future is getting nearer, and faster than before.'

'Are you optimistic?'

David nodded and smiled, 'Surprisingly, yes. Partly because *The Endless Storm* keeps selling. I get many, many young people writing to me saying how much they care about this world, and how worried they are after reading that book. The fact they care so much gives me a lot of hope for the future.'

'Your second post-Wright book, *This Poisoned Land*, led to many questions being asked about pollution, especially of our water supplies. Were you happy about that?'

'Obviously yes. It became clear when I was researching the story that too many companies were dumping their waste down the toilet or the sink, and not treating it properly. The book showed an extreme example, but we have seen many of our rivers polluted and all the fish die. The Thames, for example, is a dead river as it passes through London, which is very sad. We were in Stockholm last year and saw people fishing right in the city centre, catching salmon next to their parliament building. Wouldn't that be wonderful in London?'

Melvyn Bragg nodded and said, 'It would. Well, we must take a short commercial break. Back soon.'

An advertisement for Fairy washing up liquid started and Will turned and kissed his boyfriend. 'You looked good. Relaxed and calm.'

'I sound very posh.'

'You are very posh.'

'I don't feel it. The garden looked lovely, didn't it?'

'It did.'

'Nice job.'

Will grinned and cuddled up as the final advert, for the new Ford Escort Mark 3, finished, and the South Bank Show started again. Bragg said, 'Your last book, the third looking at the dangers of modern technology, *The Perfect Cure* came out in 1977 and was highly controversial. Both because of its topic, the drug industry, and also because your main character was gay. Why did you choose to make him gay?'

'Over the last few years, more and more people, mainly men, have been more open about their sexuality, and society at large has begun to realise just how many gay people they share their lives with. Doctors, firemen, teachers, architects, shop assistants, even some politicians. Authors too. We are everywhere. I felt it was right to make a main character gay at this time, but I wanted to write the character in such a matter-of-fact way that his lifestyle didn't impact his dedication to the truth in his fight to get justice for the patients damaged by the drug in question.'

'You are a gay writer?'

'Not in the sense of James Baldwin or Edmund White. I don't write gay literature. Well, I don't write literature. I write stories. There's a difference. However, I am a writer who happens to be gay.'

'This is the first time you have discussed this. Why is that?'

David sighed and said, 'I'll be fifty next year and for well over half my life, being gay was both illegal and viewed very negatively by society. Still is, by some. We learnt to keep our feelings and our identities hidden if we wanted to avoid personal humiliation, and even prison. It is hard, if you have grown up in such a society, to start being more open about it, but again, younger people inspire me so much with their strength and bravery, marching in public in Pride parades and fighting for equality. That's one reason. The other is it is irrelevant for the most part when it comes to the stories I have wanted to tell.'

'Is it easier now, being openly gay?'

'Yes and no. I was sitting on a wall in Padstow down in Cornwall a couple of weeks ago, with Will, my partner, and I had my arm around his shoulder as we were eating ice creams. Such a normal thing for a happy, loving couple to do. But two skinheads walked by us and shouted out "Faggots" then ran off laughing. So unnecessary, so hurtful.'

'What did you do?'

'I turned to Will and said, "One of the problems with being gay is you come across a lot of arseholes…"'

Bragg had almost died.

Back at Wisteria House, David laughed and said, 'Oh good, they kept that in.'

Bragg recovered and went on, 'Has the book been criticised because of the gay character?'

'Not really. I got one or two letters from readers saying they loved the book but hated that aspect of it, and, of course, there was John DeVoite's statement in parliament...'

Bragg nodded, 'The MP who stood up and said your book should have a warning on the cover about the gay character so as to not shock people. He said he was disgusted...'

'Yes. That was three weeks before he was caught having sex with a schoolboy in a public toilet in Camden. Typical Tory hypocrisy at its finest.'

Melvyn Bragg smiled and went on, 'The other controversial issue over *The Perfect Cure* was the poor testing of new drugs and their unintentional side effects on patients. Was it based on the Thalidomide case?'

'Partly yes. And the fight by the victims for better compensation. The Sunday Times is doing an amazing job of keeping the issue alive. And, since that drug was banned in 1961, the rules for testing have changed, but the pharmaceutical companies have shifted their research facilities to poorer countries without good health regulations, where side-affects can be hidden away from public scrutiny, and the damage brushed under the carpet. I wanted to expose that.'

Bragg nodded then said, 'I was struck by one line in particular. When the CEO of the American company says that the ideal product for them is a drug someone needs from birth to death, that keeps them alive for seventy years or more. Not a cure, but a therapy that does just enough to stop them dying.'

'Yes, I liked that line too. But I hope the industry views it as a warning, not a blueprint.'

'Some say you are anti-business.'

David laughed and shook his head. 'Absolutely not. We need business. For jobs, for success and to take more people out of poverty. And if it wasn't for business, I wouldn't be published or printed or sold around the world. But business must be honest, and if it makes mistakes it must own up and take responsibility. The Thalidomide victims should not be having to fight and fight and fight for justice. Their lives are tough enough anyway, without that.'

'Do you think it could have been avoided?'

'The whole problem? Perhaps. One of the issues with new ideas and new technologies is we rush them into production without giving much thought to what could go wrong. One of the benefits of being a writer is that I can sit here for months thinking just such thoughts. I do my research then look at the results and think, okay then, what problems could this cause?'

Bragg nodded and said, 'One of the reasons your books are so popular is that detailed research. I was talking to a couple of investigative reporters about interviewing you and they both said your books were like tutorials for journalists as to how to find information.'

David grinned, 'I'm delighted to hear that. I think it is important to be accurate when writing about a subject. It makes the book more realistic and genuine. It is partly why it took me nearly six years to write this last book.'

'Fredrick Forsyth said he followed your lead when he wrote *The Day of the Jackal*.'

That bestseller told the story of an assassination attempt on Charles de Gaulle, the French President, back in the sixties.

'Then I am very flattered. I think Jackal is the finest investigative crime novel of the last two decades. Brilliant in its detail.'

'And your next book?'

'I can't say, but it looks at another common item in everyday life that I think could cause problems in the future.'

Bragg nodded thoughtfully then asked, 'Do you like technology?'

'Oh yes. I mean, we talked about special effects in films. I love new things, like VHS recorders and computers. I used to write all my books in longhand on paper, but now I use an Apple II, which is an amazing machine. I think we will see such computers playing a huge role in our lives in the future.'

'You like the Macintosh?'

'I do. So much so I'd love to buy shares in the company if it ever goes public. It's going to be very big one day.'

'We must take another break. Back soon.'

Will grinned, 'You've outed us now?'

'I've outed Connor Lord. They have known about us in the village for years. Do you mind?'

'Of course not. I was never really ashamed of being gay, just being caught in the old days. I'm proud of you for saying it.'

'Really?'

Will kissed him and said, 'Yeah. You're my hero.'

'And you are the love of my life.'

'Would you marry me if we could get married?'

'In a heartbeat. I've practiced getting down on my knees in front of you for years, haven't I?'

Will giggled, 'Indeed you have. Oh, here we go again…'

They were back in the sun-filled garden.

Bragg said, 'We touched on politics, or rather a politician, in that last segment. Are you very political?'

'I'm very interested in politics but would never stand for public office. I think I can raise issues better through my writing than I could as a councillor or a member of parliament.'

'What do you think of Margaret Thatcher?'

David smiled and shrugged his shoulders, as if resigned to a problem.

'Well, I always thought, if we were to have a woman Prime Minister, it would be a Labour one first. Barbara Castle would have been amazing in the role. I am glad we have one, it is long overdue. However, Mrs Thatcher went into Number 10 quoting St Francis of Assisi if you remember. "*Where there is discord, let me bring union. Where there is error, let me bring truth. Where there is doubt, let me bring faith.*" Since then, she has seemed determined to bring discord wherever she goes. The demonization of the poor and unemployed is a classic example. She seems to think that they are all feckless and lazy. That is both wrong and very unkind.

Most unemployed people are desperate for jobs or have serious mental or physical health issues which prevent them finding work as companies don't want to employ them. Such people are not helped by being called scroungers. She seems to have forgotten that, once elected, a Prime Minister has to be the leader for the whole country, not just the people who voted for her.'

'I see. Finally, you might not have given many interviews over the years, but you are well known for encouraging younger authors. Many have said how much they have appreciated your words of wisdom when they have written to you, asking about writing.'

David laughed and replied, 'I'm not sure how wise my words have been, but I always try and answer my letters personally. If someone takes the trouble to write, the least I can do is reply. And give advice if someone asks for it. It is very tough to get published nowadays.

Time and time again you see, on the covers of new books, "This book is just like that book". In other words, publishers don't like taking risks; they want to publish something they know will sell, so they compare it to something that has already sold. New writers, with new original ideas, have a hard time getting into print. But I have recommended a few to contacts in publishing and asked them to take a risk. Two or three have done well as a result, but that is due to the power of the new authors' writing, not my help.'

Bragg smiled and said, 'That is very modest of you, Connor. Thank you for talking to me for *The South Bank Show*.'

'My pleasure, Melvyn.'

David got up and turned off the TV.

'Well? What do you think?'

'I think I live with a very handsome, talented, sexy man. But then I always have.'

The telephone rang. It was Gloria Hilton, who, since her husband's death two years before, had taken over the publishing company that produced his books. 'David! Darling, that was wonderful.'

'Still want to publish me?'

'Of course. You looked great and came over as the lovely man I've known so long. Well done.'

'Thanks, Gloria.'

'How's the new book coming along?'

'You'll have the first draft this time next year.'

'Right on schedule. That's what I love about you. You are so reliable. Love to Will.'

She rang off. Considering she had had to wait so long for *The Perfect Cure* and had hassled him almost daily for five years to get the manuscript, he wasn't sure she meant what she said about his reliability. David fetched the wine bottle and topped up both their glasses. Henry came in from the kitchen, tail wagging and looked at him as if trying to communicate telepathically. It worked. 'Do you want to go out?'

More tail wags, and a soft woof, so David opened the French windows and the dog trotted out into the garden. It had finally stopped raining. David and Will followed him, watching him sniff around the borders before weeing against a lavender bush. David said, 'Your turn next. When's your programme on?'

'Thursday night.'

The director of *The South Bank Show* had been very impressed with the garden that early June and had told a friend of his who produced the BBC's *Gardeners' World* programme. A month later, they had arrived and filmed Will talking about the borders, and plants that attracted insects. It was something new for the venerable TV show, and the producer had been very impressed with Will's performance. He had been provisionally booked for further work the following spring.

'We'll set the video for that too.'

'Hmmm. I haven't seen what they filmed. I hope I don't sound too stupid and thick.'

'You never sound stupid and you're thick in all the right places…'

Will ran his hand over David's crotch and said, 'You too.'

Chapter Eighteen

If you had asked the occupants of Wisteria House what they considered the most important innovation that had been developed in the last few years, David might have said the Apple computer, but Will would have chosen the VHS video recorder. They bought one on the 24th of September 1980, because of *Doctor Who*.

Will had never missed an episode since the series had started back in November 1963. He loved all science fiction, including *Star Trek* and the two *Star Wars* films, but the Time Lord's adventures were his favourite. The reason they bought a video recorder then was because, for the first time in seventeen years, he was going to miss an episode due to a social event. What was even worse was that it was his social event, not David's.

The Stroud Horticultural Society was holding its annual dinner dance, with a lecture before the meal by the venerable TV gardener, Percy Thrower, at the exact same time as the Doctor's episode was on. Will had been a member of the society for several years by then and was a fan of the famous TV gardener. He couldn't miss his talk.

He had been grumbling about the conflict, and missing the programme, for weeks when David said, 'Why don't we buy one of those video recorder things?'

'Seriously? They are very expensive.'

'We're not exactly short of cash, love, are we?'

'Um, no. Really? Could we get one?'

'Yes.'

David rang the television shop in Stroud that had supplied their latest Sony Trinitron colour TV and they delivered the video recorder on the Friday before the dinner dance.

Wisteria House was the first home to get one in Fletcher's Cross, and the man showed them very carefully how to use it so they would not set it up to record the other channels by mistake. They tested it by recording bits of programmes that night then, on Saturday afternoon, set it up to record the Time Lord and went off to Stroud.

They got back at ten thirty and, as David let Henry out for a wee, Will rewound the tape and clicked play. The famous theme music started and he was watching his favourite programme five hours after it had been sent out by the BBC. They watched the first episode of *Meglos*, a story about an intelligent cactus, together then Will rewound the tape and watched it again. And again. David went to bed.

The arrival of the video recorder had a profound effect on their lives, and the same impact on families across the country. Up until that point, if someone wanted to watch a programme, or follow a series, they had to be sitting down in front of their TVs at the time it was shown. There was no other way to watch. Now they could see it at the time of their choosing and go out and do something else at the same time it was being broadcast. They could even record one programme and watch another. Suddenly everyone who could afford a video recorder had freedom from the TV schedules.

Over breakfast on Sunday, Will went on and on about how cool it was to watch *Doctor Who* over and over again. David laughed and said something he would occasionally regret later on.

'You should record all the episodes on tape, then you can watch them again when it is off air, during the summer for example.'

'Whoo, what a great idea. I hadn't thought of that. Oh yes, that would be amazing.'

Will ended up building a huge collection of tapes, not only of *Doctor Who*, but gardening programmes, *Star Trek*, and many other shows. The TV table was replaced by a long cupboard, stacked full of his tapes, with more stored in cardboard boxes under the stairs. For a while, he spent more money on tapes that clothes.

The video recorder later brought another innovation to their home, gay pornography. They had never been a couple who bought magazines or gay films on reels to show on portable film projectors, but now VHS tapes of American porn films began to circulate. Paul Ford gave them their first one, a pirate copy of a film called *Sailor in the Wind* by an American director, William Higgins.

In a dusty pine forest, and in the living room of a ranch, various very muscular men had sweaty sex right before their eyes in Wisteria House. It was a huge turn-on and they ended up having sex in front of the TV, much to Henry's disgust.

Another one introduced them to leather sex. They had been watching a film about a master and a slave, and David had noticed how very, very excited it had made Will. He had contacted Paul and asked if he knew of any shops that sold leather clothing, and his editor had recommended a tiny, back-street place near Spittlefields meat market.

David had gone there during a solo visit to town, and purchased a leather waist coat, the item that had attracted Will's attention, and a pair of handcuffs. That night, when Will was cleaning his teeth, naked in the bathroom, he had looked in the mirror and seen, standing behind him, leaning against the door lintel, his lover wearing nothing but this item of clothing, holding out the silver cuffs. A very excited lover.

He had gasped, turned and then dropped to his knees. They only played with the waistcoat every now and then afterwards, but it was always fun when they did.

They didn't get addicted to gay porn or S&M but enjoyed getting hold of new films every now and then, to spice things up a little. Paul was an avid enthusiast, and often gave them copies of films he had got hold of, and they bought a few original copies as well, from a sex shop in Soho that soon realised gay men loved such films, especially as some of the pirate copies were very grainy and poor quality.

As the original copies had colour pictures of naked men on their covers, they stored them in a separate cardboard box under the stairs with "Old Invoices" written on the top. They didn't want Lottie, or Charlie, to find them.

*

If videos became Will's new passion, cooking had become David's. He had always enjoyed preparing their meals but had realised that he was a bit stuck in a rut when it came to the variety of dishes he cooked.

He started buying cookbooks, covering a range of cuisines, including the Le Cordon Bleu Cookbook, Graham Kerr's

The Galloping Gourmet and, of course, Delia Smith's books, tie-ins to her first BBC cookery programmes.

He started baking different types of bread, making pastry for wonderful apple pies and sausage rolls, and extending his range of meals to cover world cookery too.

He started making Asian food, and asked Will to grow coriander and basil in the greenhouse and daikon, white radishes, in the walled garden, as it was hard at that time to find them in the local shops. Then he asked for fennel and artichokes as well.

The gardener loved the challenge of growing new plants, even if he didn't always enjoy eating them. He grew two types of celery for David, but never ate the stuff himself. David loved stilton with a good, hearty stick of celery, or smothering a stalk in Philadelphia cream cheese to eat with a sherry before Sunday lunch.

They started to eat a lot more rice, and he began making curries as well.

On a trip to Italy, he had purchased a pasta making machine and, after a few failed attempts, became an expert at making their own fresh pasta, a real novelty back in the late '70s and early 1980s.

He tried pickling herrings, after their visit to Stockholm, but Will hated them. However, he started growing dill, and David's "lamb in a creamy dill sauce" became a firm favourite. He also started using anchovies to add flavour to lamb stews and his more traditional meals.

He made most of their food from scratch, but there were three things he always had to buy. One was watercress, which they both loved. Will could not grow this in the garden as they didn't have a stream with flowing water. A local farmer's wife started growing some, and sold it in the first, very small, farm shop in the area. She also raised Aylesbury ducks, which they roasted and served with their own apple sauce. The second thing they had to buy was Melton Mowbray pork pies, another favourite. David tried baking one of these but Henry had to eat it as it was a total disaster. He also had to buy cheese, including a very mature cheddar from a local dairy that they loved.

Will, who had been raised on baked beans on toast or fried spam with a fried egg, adapted to these new food experiences happily. The only problem arose when David was trying something for the first time and got it wrong.

His initial attempt at a curry was so hot and powerful it reduced the two men to hysterics as they fought over the water jug.

When it came to alcohol, they didn't drink that much. Will would still go to the pub on his own once a week and sit with his old mates, now all family men escaping the clutches of their wives and children. There he drank a few beers or a cider. At home, they drank cider or wine with their meals, and took the occasional single glass of whisky before going to bed.

Unlike many wealthy people, they didn't collect wine, or spend that much on expensive bottles. A good red or white was more than enough for them. They didn't need bottles that cost hundreds of pounds to enjoy a glass with their meals.

They had dinner parties. John Ridge and his wife would drop around now and then for a meal; unfortunately, Ridge had been elected the secretary of the local golf club and had become a huge golf bore, which they both found very tedious, so their visits became rarer and rarer. They had weekend visitors from London; Murrey Holt and his wife came to stay, as did Paul Ford and his partner. David would prepare elaborate menus for their guests, leaving them bloated but happy as they returned to the big city on Sunday nights.

Charlie Nolan also loved David's food, especially his big quiche Loraines. These would be cooked, then left, cut into slices in the tin cooking form, in the fridge, where anyone could help themselves if they were feeling hungry. There would often be a plate of cold sausages as well, and those never lasted very long if the lad was about. However, from the look on Henry's face and his tendency to lick his chops after a visit from the boy, David and Will realised he was feeding their dog rather than enjoying the sausages himself. Lottie and Charlie would often stay from supper during his school holidays. She was no longer just the cleaning woman, but a firm friend and part of their lives, as was her grandson.

*

Even though they lived in a quiet village in the countryside, they kept in touch with what was going on in the world through newspapers, the radio and, of course, the evening news.

Although Will was addicted to television, they had both sat down in

1978 to listen to a new science fiction series on the radio called *The Hitchhiker's Guide to the Galaxy.*

They had become huge fans of Douglas Adams' comedy, quoting its phrases like "Don't Panic" endlessly for a while. David thought the famous answer to the meaning of life, the universe and everything, which was 42, was a brilliant idea, and answer. Will just liked the idea of a computer called Deep Thought and loved Marvin the paranoid android.

But the television was their main source of news. They had followed all the major events during the seventies; the industrial unrest that finished off the last Labour government; the Queen's Silver Jubilee celebrations, Thatcher's arrival at Number 10 as the first woman Prime Minister, and the other big political stories of the decade as well.

When Iran had gone through the Islamic Revolution in 1979, they had both been shocked, as they had visited Tehran in 1964, doing research for David's ninth John Wright book, *The Tehran Theorem.* They had found a very westernised society, with beautiful women wearing the latest fashions, bars and cafés where alcohol flowed, and very warm and friendly locals. They knew that the Shah was probably not the nicest man in the world, but the capital was like any other major world city, like Paris or Rome.

When Ayatollah Khomeini returned and the country became an Islamic republic, David had been stunned. Even his vivid imagination had not considered a country returning to be ruled by a religious government, that imposed strict Islamic rules of its citizens. As time went by and they saw images of women, their bodies fully covered, and the wild protests against America, he briefly considered writing a novel about it.

However, seeing the fanaticism of some of the Ayatollah's followers, he felt it might not be the cleverest idea he had ever had. When Salman Rushdie had a fatwa declared against him for writing *The Satanic Verses* by the Ayatollah, he was, once again, proved right.

They watched all those events live. It was the only way one could. When the video recorder arrived, David started taping news programmes for future reference but tended to forget to watch them later.

Eventually Will would record over his cassettes, adding further adventures of *Doctor Who* to his growing collection.

The machine was still a new toy when they recorded David's interview, but it would become an integral part of their lives from then on.

Chapter Nineteen

Lottie arrived at eight on the Monday morning after "The South Bank Show", telling them how much she had enjoyed the interview.

'It was funny seeing you on TV. You looked so natural.'

'Thank you, Lottie. But I'm not going to make a habit of it.'

'And what you said about that bitch Thatcher. I cheered.'

David laughed. 'I bet she didn't, but I doubt if she watches such programmes. How's Charlie doing? Did he win on Saturday?'

Charlie, now ten years old, had just gone back to his prep school in Combe Weston. He went as a day boy and had just started playing rugby. When he had first arrived, he had been bullied because of the colour of his skin, but David had suggested that he focus on games, as all boys liked a winner. He had excelled at football and cricket and had now started playing rugby.

'Oh, he did so well. Scored three tries. He's ever so popular now.'

'I bet. But if he gets teased again, let me know and I'll go over and see that headmaster again.'

Charlie was ahead of other children his age as, once he was old enough, David had got him reading and bought loads of children's books. When it came to sports, Will had cut out an area in the long grass meadow behind the magnolia tree so they had a football field and put up some goal posts.

Both of them would have kick-abouts with the lad, keeping them fit as he developed his skills. He helped Will in the garden from time to time but preferred looking at the fish, frogs and newts in the pond to weeding the borders.

Charlie also had a beautiful singing voice, just like Flo's.

He was now a member of the church choir and had stunned the

congregation the previous Christmas with his solo of "Once in Royal David's City". There hadn't been a dry eye in the church when he finished.

When he had been bullied that first year, David had driven over to the school and given the headmaster hell. The man had not been used to parents challenging him in his study, but David made it very clear he would be monitoring the situation and would write to the Department of Education if things didn't improve.

Charlie had settled down and, as predicted, once his sporting skills were demonstrated, had become a popular boy. He was also bright, helped by the early reading, and did well with his lessons. They hoped he would get into Stroud Grammar School, as Lottie didn't like the idea of him going to a private school for the rest of his education.

'He's Flo's lad, not some posh boy.'

*

Will got up from the table that Monday and said, 'Right, I must get back to the garden. Got to clear up the vegetable patch for the winter. There are still a few leeks left. Do you want them for dinner, love?'

David nodded. 'Yep, bring them in. I'll get some lamb chops. Or do you want them, Lottie?'

'Not leeks, but have you got any tomatoes left? Charlie does like his toms.'

'Yes, got a lot of those too. I'll get you a pound or so.'

'Lovely. Now, I must get on.' She left the kitchen to inspect the house and see what needed her attention first. David sat, scratching Henry's head, planning the day's research. The back door opened and a woman's voice called out, 'Morning. Is there anyone here?'

'You sound like a medium, Ginny.'

'Oh, hello, TV star. You were super last night.'

Ginny Rose swept in, all blond hair and bangles. She was thirty-five, lived on the "new" estate, was married with two girls at the primary school, and the latest addition to Wisteria House. She was David's secretary and assistant.

He had always tried to manage his papers and correspondence by himself, but, in recent years, the sheer number of letters, and the volume of past correspondence and research notes, had begun to overwhelm him. He had advertised for help and Ginny had arrived the year before, having moved to Fletcher's Cross from London. Her husband, Bill, was a solicitor in Stroud, and she had been the private secretary to the CEO of an oil company in the City, before the birth of their children.

On her first day of work, he had taken her up into the attic and shown her his collection of banana boxes, filled with his original manuscripts, his research notes and all his old letters. He never threw anything away, so there were nearly a hundred boxes up there. She almost resigned on the spot, but Ginny was made of stronger stuff, and she had taken charge of David's vast archive and, in just over six months, had brought order to his world.

The only embarrassing moment in their relationship so far had been when she had found a box marked "Old Invoices" under the stairs…

With the archive under control and the daily flow of letters more manageable, she had started helping him with research too.

David smiled at her that Monday morning and said, 'Glad you approve. Now, we need to try and sort out all that legal crap from the States today.'

'Oh God. There is so much of it…'

The new book, called *Smoking Gun*, was about the cigarette industry. Through a fictional trial, a class action suit, David was trying to tell the story behind cigarette companies covering up the dangers of nicotine.

Although the dangers had been known for twenty years, no one had yet successfully sued Big Tobacco, and cigarettes were still available without health warnings in England. His story included politicians again, on both sides of the Atlantic, tobacco growers, state tax authorities, smugglers, the black market and individuals suffering from the effects of long-term smoking. Since he had started the book, he and Will had given up cigarettes, so shocked had he been as his research revealed the true cost of smoking.

He had advertised in newspapers, in America and Britain, anonymously, for whistle-blowers to send information to a post office box in Stroud about court cases or deceptions by the tobacco companies, and material had flooded in. From American lawyers, from ex-employees of Big Tobacco, from law makers; they now had several box files full of papers to sort through.

He got up and the two went into the study. It would be a very long day.

*

Out in the garden, Will also had an assistant. A seventeen-year-old lad called Callum Sheare, who he had caught trying to steal one of the chickens back in March. Rather than call the police, Will had sat the boy down in the potting shed, made him a mug of tea using the electric kettle
he had there and said, 'Come on, what's going on then?'
The boy had been dressed in a biker's leather jacket over a dirty white T-shirt, torn jeans and a pair of very scuffed-up Doc Marten's black boots. Rather than acting tough and stroppy, he had been very nervous; he was skinny, and had a bad case of acne. He had looked at Will with uncertainty and a definite look of suspicion, as Cal was unused to someone being kind, and he knew the occupants of Wisteria House were gay. But over a mug of tea and some biscuits, he had poured out his story.
He lived in one of the council houses outside Combe Weston with his single mother, who was an alcoholic. His dad, a farm worker, had left them when Cal was eight, going off with a woman from Stroud. He hadn't kept in touch.
Cal had done badly at school, written off as a lazy little bastard by his teachers, who hadn't realised the boy was very dyslexic. He had bunked off from lessons often and left at sixteen with no qualifications whatsoever, and was trying to survive by poaching, stealing anything that wasn't tied down, and shop lifting. He did all this to look after his mother; hence stealing the chicken. He hadn't been caught by the police yet, but Will could tell that it was only a matter of time. Even if he didn't call them himself.
Will had listened to this miserable saga patiently but had to stop as he needed to double dig the vegetable beds ready for the spring planting. He suggested Cal help him and offered to pay him if he did. They worked side-by-side for the rest of that day, and Cal reappeared the next day, and the day after that. He needed to be taught everything about plants and gardening, but only needed to be told once. He was a natural.
By the end of the summer of 1980, he was working full time, had a girlfriend in the village, and had stopped stealing from shops;

he still did a bit of poaching, but that was mainly to get one over on the new owner of Langham House.

Lord Langham had died, leaving terrible debts, and his widow had had to sell up. The girls had all finally found husbands, but Susan Langham had had to leave the area and now lived in a tiny flat in Weymouth, on the south coast. The estate had been bought by a millionaire called Huxton who ran a chain of second-hand car dealerships across the country and who prided himself on his new country gentleman's life, including raising pheasants.

Cal had a running battle with Huxton's gamekeeper, who constantly tried to stop him, but the lad was careful and sly, and often managed to bring home a plump bird or two. David and Will had enjoyed more than one Langham pheasant that summer.

The two gardeners were out in the apple orchard, harvesting the last of that year's crop. Cal had also watched the programme and was saying, 'I never think of you two being poofters. You're not camp like them on the telly.'

'Why, thank you, kind sir. And the accepted term is gay, not poofters.'

'Right, sorry. But you're not, are you? Camp?'

'Never felt the need to be. And old guys like us, as David said, we had to hide who we were back in the day, so being effeminate just wasn't on the cards. Not round here, anyway.'

'You're not old.'

'Ta.'

He moved the first basket to one side and slid the empty one in its place, ready to receive the next batch of fruit. They had already harvested hundreds of pounds of apples and Lottie and David had made loads of apple sauce for the house, as well as supplying the Fish and Trumpet with fruit for the last month. The pub now did Sunday lunches, and their apple pies were very popular. The apples they were collecting that day would go onto wooden racks in the shed, to last them through the winter.

'How's Nell doing? Enjoying working at the Frog?'

The Fletcher's Arms had closed down and the building had been bought by a local businessman and turned into a restaurant. It was doing very well, and Cal's girlfriend had just started as a waitress there.

'Yeah. She does. But some of them punters. They take liberties with her. Slapping her bum like. I don't like that, but she says it is all part of the job. They tip better if she don't complain.'

'I think that's really rude. A woman should not have to be molested to have a job.'

'Right. Well, there's no other work around here at the moment. I've told her to tell me if it gets too much and I'll have a word with Arty.'

Arthur Jackson, or Arty to his team, was the young chef who ran the kitchen. The food at the Frog was excellent, and it was starting to gain a reputation as a destination dining spot. The neighbours, glad the bikers had gone, were now complaining about the parking at night, as more and more people came from Stroud and the surrounding area to eat there.

Cal said, 'When's your show on?'

'Thursday.'

'I'll watch that then. Fancy me working with two telly stars.'

'We are hardly that. I might be on for about five minutes.'

'But they want you to do more next year?'

'They said so, but nothing's booked for certain. I guess they want to see the reaction to this week's programme.'

'Bet they'll like you. You're good at explaining things. You helped me.'

'You're bloody good at what you do, mate.'

The boy looked very pleased. He was still very unused to praise, a little like Will had been when he had arrived at Wisteria House.

They carried the two full baskets over to the big wheelbarrow and walked back to the shed to store them. Henry was waiting for them. He knew when they had

their elevenses, and also knew that, if he was patient, he might get a biscuit. Cal gave him a hug and said, 'You greedy bugger. You wants a bickie, right?'

Henry barked happily. He was now nearly twelve, but still acted like a huge puppy. He trotted into the shed and sat waiting as Will clicked the electric kettle on, plonked two tea bags into mugs and Cal fetched the tin.

'Give him one. Only one. He's getting fat.'

Cal slipped him two, then sat down and started rolling a cigarette. 'You want one?'

'No thanks. After hearing about David's research into smoking, I'll never touch one again.'

147

'That's what his new book's about then? Smoking?'
'Yes, but don't tell anyone. It's a surprise.'
'Ok. Now, what's next?'

Chapter Twenty

That night, after a home-made lasagne and half a bottle of red, David and Will were laying, side by side, on the sofa. They were both tired; Will physically, after a long day's gardening; David, his head spinning from the details about the legal machinations of the American tobacco companies he and Ginny had read through. They both needed to relax and cuddling up together peacefully was still one of their favourite ways of doing that. They often didn't speak much when they did, but just enjoyed the closeness and body contact. David stroked Will's hair, still thick and blond. His own was now peppered with grey hairs, but they made him look distinguished, not old.

'This is nice.'

David grinned. 'It is. Always is.'

'Two old poofters on a sofa.'

'Poofters?'

'That's what Cal called us. I put him right.'

'Good. Is he still doing okay?'

'Yep. Don't know how I managed on my own all those years.'

'Keeps you fit though.'

He slipped a hand under Will's T-shirt and stroked his firm, hard belly.

'Hmm. Still want me then?'

'Always. That will never change.'

'And only me?'

'Yes, that won't change either.'

'No lingering regrets about New York then?'

'There never were, you know that.'

Will was referring to their one and only trip to the Big Apple the previous year. David's American publishers had pushed the boat out for the visit, putting them up in the Presidential Suite at the Waldorf Astoria, and providing them with a guide, a PR guy called Alex, whom Cal would have easily realised was gay, such was his very effeminate way of acting and talking. Alex had taken them everywhere, and shown them the sights, including a helicopter ride around the Statue of Liberty. He had also gotten them invited to a very special party.

It had been the gay social event of the year, the birthday party of Tony Fleck, the fashion designer, at his vast brownstone duplex overlooking Central Park. They had arrived at nine and found themselves in another world. The guests were all in dinner jackets (Alex had rented them tuxedos), but it was the serving team and the "extras" who had taken the two visitors' breath away. Every waiter had just been wearing tiny black satin shorts and bow ties, and they were all incredibly beautiful young men, with buffed bodies.

The extras, also extremely fit young males, included two naked black guys, one endowed with such a massive penis they thought it was artificial, who wandered around the exclusively male guests, encouraging them to touch and play with his huge organ.

The apartment was hideous. A mix of mock-Versailles and Las Vegas, a nightmare of golden painted chairs and tables, gold and white pillars, and vast paintings of their host dressed as various French kings, with wigs and jewellery dripping off every fat finger. A string quartet was playing Elton John songs very badly.

The waiters were walking around with trays offering glasses of champagne, ready rolled joints, piles of "uppers" and "downers" and silver platters with lines of cocaine and gold tubes for snorting it. Some even had canapés, but the guests seemed to ignore those. David and Will recognised a few well-known faces.

A Hollywood actor, famous for macho roles in Westerns, had his arm draped around the naked chest of a very young teenage boy, who was wearing nothing but a white Stetson and a jock strap; a vastly fat Republican Senator was sitting on a sofa bookended by two well-endowed leathermen in chaps and harnesses, and a very well-known Christian television evangelist stood by one wall with his hand between the arse cheeks of a naked Asian muscle man.

He looked as if he would soon be begging His Lord for forgiveness for months afterwards.

Alex had introduced them to their host, who had greeted them regally but without much enthusiasm, before clapping his hands and sitting down on a golden throne. In front of the throne a circular bed had been placed in the main living area. The guests had formed circle around this and the two black men had had sex in front of the very excited crowd. As the guests whooped and hooted their encouragement, the other extras started leading guests off into bedrooms or bathrooms, and it became clear very quickly that this whole party was turning into an orgy.

David and Will had never been to such an event, and they still had not had sex with anyone else since they had got together. David was surprised by the scene on the bed, as Alex hadn't warned them it was that sort of party. He saw Will go bright red and head out onto the vast balcony that ran along the front of the apartment. He followed him and found his lover leaning against the balustrade, gasping for breath.

'Hey, relax, come on, sweetheart, sit down.'

'I can't breathe. God, I wasn't expecting...'

He sat down on a wrought iron chair and put his head between his knees.

David gently rubbed the back of his neck and said, 'That's it. Deep slow breaths. Take it easy.'

'It's so... crude. All those men just cheering them on.'

'I know. One hears of these things but seeing them...'

'Do you want to fuck any of those guys?'

'God no. They are prostitutes. And I don't want to fuck anyone but you anyway.'

'Are you sure? Some of them are so beautiful.'

Will was still gasping but his breathing was becoming steadier and calmer. He looked up at David with concern and worry in his eyes. His boyfriend replied, 'You are the only man I have ever wanted and ever will.'

'Really?'

'Yes. I think we should leave.'

Will looked relieved, 'Can we? I hate this. But I don't want to be boring.'

'It's not boring. This scene... it's just not us, is it?'

Will looked very relieved as he said, 'If you're happy to go, I'd love that.'

David pulled him to his feet and hugged him.

'Come on then. We'll slip away and find a taxi back to the hotel.'

They made their way through the crowd. The black couple had now dragged two guests onto the bed to join them and several people were having oral sex against the walls of the living room and in the hallway. David recognised one of the new men on the bed, their guide, Alex. He screamed in delight as he was impaled by the huge black man, the crowd cheering them on. The two Englishmen reached the front door and went down in the elevator and were lucky enough to catch a yellow cab just as another guest arrived. He looked at them as if they were mad to be leaving such a wonderful event.

Back at the hotel, David stopped at the reception and ordered a bottle of Famous Grouse whiskey and some cold beef sandwiches to be brought up to their suite, then they went up to their rooms. Will took off his bow tie and jacket and went and washed his face. David took off his tie, undid his collar and stood in the lounge area, still stunned by what they had seen.

A knock on the door announced the arrival of their order and the waiter laid out two plates and a platter of sandwiches, two tumblers, an ice bucket and a bottle of scotch on the huge onyx coffee table. He left with a ten-dollar tip. Will came out of the bathroom and sat on the sofa as David poured two whiskeys, added ice, then sat next to him, one arm around his lover's shoulders. They clinked glasses and David said, 'Cheers. Glad we're out of that.'

'Are you sure?'

'Yes. I thought we might meet a few nice new guys if we went. I wasn't expecting an orgy, nor did I want to take part in one.'

'We're very boring.'

'We're very faithful, and I see nothing wrong with that. I have never felt the need for a bit on the side. You are more than enough for me.'

'I feel the same about you. Always have.'

'Then that's that.'

They drank silently, then Will started laughing.

'What's so funny?'

'Us. Like two old Victorian ladies, shocked by modern society.'

David nodded. 'Well, I wasn't shocked, just a bit disgusted. And all those drugs too. Way over the top.'

Will reached for a sandwich. 'We didn't eat anything.'

'Not sure I wanted to. God knows what was in those nibbles.'

'LSD, probably.'

'Probably.'

'Was that Alex on the bed when we left?'

'Yep.'

'He'll be sore in the morning.'

'If he's still alive. That was a very lethal weapon.'

'Not even John Wright could have fought back against that…'

*

They hadn't seen Alex again. The next morning, they went shopping for clothes, and a model of the Millennium Falcon spaceship for Charlie, who was a huge *Star Wars* fan. At four that same day, the limousine from the publishers collected them and drove then out to John F Kennedy airport and their flight home, first class, on a BA Jumbo jet. They arrived at Heathrow, collected the Range Rover, their current car, from the car park and drove back to Fletcher's Cross, still shell-shocked by the sights and sounds of New York City.

Back on the sofa, Will kissed David on the lips and sighed happily. 'No, orgies aren't our thing, are they?'

'Certainly not. We're British.'

'I'm sure they go on here too. In backrooms or private houses.'

'Probably. But not at Wisteria House. Henry wouldn't put up with it…'

Chapter Twenty-One

During the seventies, the Cotswolds had seen a steady increase in visitors, and a lot of interest from people wanting to buy properties there as weekend hideaways, or simply move to live and work in the area. As a result, Studley Combe, the village to the south, had been designated a place for expansion, and a further thousand homes had been built, including a council estate of three hundred flats and semi-detached houses.

A small shopping centre serviced these new citizens, with a Sainsbury supermarket, a Boot's chemist, an optician, a Clark's shoe shop, and a Homebase Garden centre.

The resulting increase in traffic between Studley Combe and Stroud led to both Fletcher's Cross and Combe Weston becoming clogged with lorries and commuters. Luckily for the occupants of Wisteria House, the main road ran along the opposite side of the village green from the house, but the number of cars and trucks had a huge impact during the weekdays; weekend day trippers only added to the problems. Demands started for a bypass.

In 1972, two children were knocked down and killed by a lorry delivering bricks to one of the housing construction sites. The children were visiting their grandmother. Their parents had parked opposite her cottage and the two little girls had run across the road to greet her without looking.

The tragic nature of the deaths, which got the full tabloid treatment, led to protests and further demands for a bypass. As the head of county planning lived in Fletcher's Cross, he pushed this through far quicker than usual and work started in 1975.

That was mainly due to Huxton being very willing to sell off some of his land for a quick profit for the new road.

Also, luckily for David and Will, the planned bypass ran on the opposite side of the village to their home.

On November 1st, 1980, the Prince of Wales opened the new bypass in a brief ceremony outside Combe Weston. The road took traffic around both villages. This would lead to a jump in house prices, as they became very quiet and peaceful again, and highly desirable.

David had anticipated this and bought another cottage, the one next to the post office. He rented this out to a couple of local artists, a husband-and-wife team, who set up a studio and pottery in the back garden.

David was concerned that the village might become too expensive for locals to live there. Already three of the cottages had been bought as weekend retreats by Londoners, and other villages in the Cotswolds were becoming empty during the weekdays, only coming alive when their remote owners arrived, exhausted by the traffic jams, late on Friday evenings. They were deserted again around five each Sunday afternoon as those weekenders headed back to the city.

But Fletcher's Cross was still a working village and a real community, and David and Will were a central part of it. They went to church each Sunday; neither was really religious, but the service was a way for the community to meet up and join together. Fallow had been a member of the "Build a bypass" committee and, despite his advancing years, was still an enthusiastic shepherd of his Fletcher's Cross flock. David had given money to help with the campaign.

The day the bypass opened, James Fallow was standing outside Wisteria House with them, watching the flow of traffic. One moment there was a steady line of cars and lorries, the next they were gone. He turned to David and Will and said, 'Well, the Prince must have cut the tape.'

'You didn't want to be there?'

'Oh, I was invited, but that's not my sort of thing. I just wanted to get the cars away from the village. Listen to that. We can hear the birds singing again.'

In fact, it was just a single seagull, calling for a partner, before it flew away to the west and the Severn estuary.

David said, 'Come on in, James, and have a cup of tea.'

They went into the living room, as Ginny offered to make them all tea. Sitting on the sofa, Fallow had looked around then fixed on two items on the mantelpiece. Even though he had been a regular visitor for many years, he hadn't noticed them before and said, 'Good Lord, are those somebody's feet?'

'They are.'

'Whose?'

'Mine,' said Will. 'It's a long story.'

'I'm all ears.'

So, David told him about Will's arrival at the house and the state of his old boots, and Fallow laughed, then frowned.

'I must say, when one thinks back to those dark days when being homosexual was a crime, one shudders at the injustice. I had two or three members of my congregation who struggled with their desires and fears. It was a very unkind time.'

'It was. But then it still is in some countries, and we haven't got equality here, not really.'

'Do you mean gay marriage?'

'Well, not just that. For example, if Will went into hospital and needed an operation, if he was unconscious or in a coma, I wouldn't have any rights to decide on his treatment. Even though we have lived together for over twenty-two years, I am not viewed as family. That's not right.'

'No. Gosh, I'd never thought of that. How dreadful. You're not ill, are you, Will?'

The gardener laughed. 'No, James, I'm fine. But it's the principal.'

Ginny arrived with a huge tray with tea things on it, including a sponge cake. She distributed cups of Typhoo and slices of cake, said her farewells for the day, then left the three of them to go and post that day's mail.

Fallow sat back, cup in hand, and said, 'Oh, by the way. I saw you on *Gardeners' World*, Will. I've been meaning to tell you for weeks. You were jolly good.'

'Thank you. I felt a bit stupid, watching myself.'

'No, no, dear boy, you were excellent. My wife's started digging up the roses to replace them with plants that attract more butterflies. Good thing too because the roe deer always come and eat the rosebuds anyway.'

'Yes, they are a problem, but Henry sees them off if they come anywhere near here.'

The dog in question rolled over and slapped his tail on the carpet, happy that his contribution to the safety of the garden plants was being acknowledged. He was still waiting for cake, but none came his way.

David sat and listened. It was so nice for Will to be the centre of attention for a change. He had received lots of letters since the programme aired, which was highly unusual. One of these was from *Country Life* magazine, which had asked if he would be interested in writing a monthly column about environmental gardening. He had shown it to David and asked, 'What do I do about this? I can't write to save my life.'

'But would you have enough ideas for such a column?'

'Oh yeah, no problem. Plant selection, garden design, types of manure or mulch, trees, the lot. I could talk about them, but not write about them.'

'So, tell me and I'll write it for you, and you send it in as if you'd written it.'

Will had gasped. 'Oh, Davey, I can't do that. That's not fair on you.'

'It would take me about an hour a month to bash out an article. You are worth an hour of my time any day of the week.'

So, Will had written back (well, David had) accepting the suggestion, and enclosing the first article, about how to start preparing in the spring for planting new insect-loving flowers, ready for the next edition. The editor had written back almost immediately, very happy with his acceptance and the high quality of the writing. When he showed this to David, the author chuckled, 'Glad to see I've still got it...'

Fallow stayed for another half hour then left them in peace. David hugged Will then said, 'I can't be bothered to cook. Shall we try and see if we can get a table at the Frog?'

'Oh yeah. Nice idea. But I'm paying. From my BBC money.'

'You're on. I'll call and see if they can fit us in.'

They could. Especially as he booked the table as Connor Lord. The Frog was very busy; they were met at the door by the charming manager, who led them to a table by the window overlooking the garden, now lit by discreet spotlights. The trees were bare, but the branches and trunks looked beautiful under the artificial lighting. The restaurant's style was French bistro meets fine dining.

The napkins were ringed by paper rolls with the restaurant's logo, a pen and ink drawing of *Rana temporaria*, the Common Frog, sitting on a lily pad, embossed on each. They were delighted to see that their waitress was Nell, Cal's girlfriend. A slim, very pretty blond girl, she was wearing the staff uniform of black dress with a white collar and a white apron.

She grinned when she gave them the menus and said, 'Oh, I'm glad it's you two. At least you won't pinch my bum.'

'You're right there, love,' said Will.

'Do you want to know the specials?'

'Go on then.'

'We've got whitebait from Torquay, fresh today. And some amazing rump and T-bone steaks. Really juicy ones. The T-bones are enormous. I think they're from an elephant.'

They laughed appreciatively. Despite the extensive menu, which mixed French and English cuisine, the specials sounded very good.

'Then that's what I'll have,' said David. 'Whitebait and a rump steak, medium rare. I'm not into elephant at the moment so I'll skip the T-bones.'

'Me too. And a bottle of Merlot,' added Will, without flinching at the price for a change. The BBC had sent him a nice cheque, the first money he had earned from anyone apart from David since they had started living together.

They looked around the restaurant. The other guests were either couples having a romantic evening out, or parties of four, talking loudly about house prices, golf or new cars. All straight. One or two glanced over at their table, then bent forward whispering, as if asking for confirmation of their identity.

When the Frog had been the Fletcher's Arms, it had been decorated in purple flock wallpaper, with a purple and orange carpet, black plastic roof beams and stank of spilt beer. Now there was a feature wall of exposed brickwork, the rest painted a tasteful terracotta from Laura Ashley, bare, stripped floorboards, tables with red and white checked tablecloths and candles stuck in old, wicker-bound Chianti bottles. It was a classic bistro look, and the aroma of warm bread and fine food had replaced the stench of stale ale.

Nell brought their first course after pouring the wine. The whitebait was delicious, with a crispy batter and was very fresh, as promised. Later, as they were finishing the excellent steaks, which were indeed very juicy, a woman came over and asked, 'Excuse me for interrupting. Are you Will Forman?'

'Yes,' said a startled Will.

'Oh, I thought so. I thought you were wonderful on *Gardeners' World*.

Just wanted to say that. Bye.'

She went back to her table, nodding enthusiastically to her friends, and they heard, "Yes, it was him. I knew it was…"

David laughed. 'It's your time now, sweetheart.'

'Do you mind?'

'God no. I thrilled for you, really, I am. I've known for years how damn good you are when it comes to gardening. You are a real expert. It's well overdue that you get more credit, and I am more than happy to share you with the rest of the country.'

Will blushed. 'You are a very nice man, have I ever told you that?'

'Often. Right back at you.'

They declined a pudding, paid up and headed out, leaving a good tip for Nell. They drove home along a very peaceful road with no traffic, the song, *The air that I breathe,* came on the radio. It was sung the Hollies.

'Oh, it's our song,' said Will. When it had come out in 1978, they had played it endlessly until, one night in bed, David had started singing it to him and Will had farted very loudly. After that they stopped playing it.

'Used to be, you windy bastard.'

'Ah, all I need is the air that I breathe and to love you, Davey…'

At home they let Henry out for a wee as soon as they had parked the car in the garage. They stood, side by side, on the terrace, hugging, as their dog inspected the borders, choosing the perfect spot for his final ablutions of the day. It was a still, cold evening, and the stars were out.

'This is surprisingly romantic all of a sudden.'

'It is, Davey.'

'Thank you for dinner. It was delicious.'

'You've paid enough for me over the years. It's about time I took you out.'

'What's mine has always been yours, you know that.'

'I know. But still, it's nice to pay my way for a change.'

'You can do it whenever you like now you're a famous gardening expert.'

'Hardly that.'
'Not yet maybe, but in the future…'

*

The final big event of 1980 for David was the initial public offering, on the New York Stock Exchange, of Apple Computers. Through his English stockbroker, he was able to buy one hundred thousand shares, for twenty-two dollars each, a total price of nearly one point six million pounds sterling at the time, his largest ever single investment. His stockbroker thought he was mad to put so much money into one company with such a limited range of products, until the share price rose seven dollars that first day, giving him a profit of over half a million pounds. He would hold onto his Apple stock for the rest of his life; with stock splits, he would end up with over five million shares.

During his time as an author, David made a lot of money from his writing, but much, much more from that single "mad" investment.

Chapter Twenty-Two

'Will?'

'Yes, Charlie?'

'What's a faggot?'

They were in the greenhouse, planting nasturtium seeds in pots, some of which would join lobelias, ivies and petunias in the hanging baskets Will put up on both side of the front door in the summer. It was Charlie's Easter school holidays in 1982. He was nearly twelve, and in his last year at the preparatory school in Combe Weston. He had passed his eleven plus exams with top marks and was going to the grammar school in Stroud in September, much to the delight of his grandmother.

'Where did you hear that word?'

'At school. A couple of boys were teasing another lad, calling him a faggot and a queer, one lunchtime.'

'What did you do?'

'I told them to leave him alone and sat with him so they couldn't bully him anymore.'

Will smiled and said, 'That was nice of you.'

'So? What are faggots and queers then?'

Will sat on the stool and looked at the lad. Charlie was tall for his age and had a mass of hair sticking up all over his head, which would need cuttingbefore he returned for the summer term. It wasn't quite an afro style, but close.

'They are both horrible, nasty words for people like Davey and me. Men who like men.'

The boy frowned. 'But you're not nasty people.'

'I know that.' Will said, laughing. 'We're great. But some people hate us because we are different.'

'Like me being black?'

'Yep. Sometimes it's the same people who hate us both. They are racists and homophobic.'

Charlie frowned, 'What's that then?'

'It describes people who hate homosexuals. Like David and me. Men who have sex with other men.'

'So, you've never been with a woman then?'

'Nope. Never have, never wanted to.'

'Why is that?'

Will shrugged. 'I don't know. Not sure anyone does. Some men like men, some like women, and they know that right from the start. Like you and Mandy. You've had the hots for her since you were five.'

'I haven't.' Charlie grinned, going a bit red.

'Oh yes you have. We've got the picture to prove it. You kissing her behind the Christmas tree in 1976.'

The boy nodded. 'She's cute.'

'She is.'

Mandy was the daughter of one of Will's old mates, Sam, still working as a farm labourer. Only now he was called an agricultural worker. Charlie pushed another seed into a flowerpot then asked, 'So why are some men like you and David?'

'It just happens, I think. It's just the way we are. Like being left-handed or ginger.'

'Are you happy?'

'Very. We've been together now for over twenty-four years. I love him very much indeed.'

'That's nice. He's great.'

'He is.'

They both laughed then Charlie looked serious. 'What about them funny buggers you used to warn me about?'

'Well, there are some men, and some women for that matter, who want to have sex with children. It is wrong, and they shouldn't, but they do. We warned you about them so you'd know it was wrong if someone tried to approach you. Has anyone?'

'No. Never. But some men like little boys then?'

'Yes.'

'That's not nice.'

'No. We certainly don't. Not like that anyway.'

162

'Good.' He thought for a moment then asked, 'Why do people hate me for being black, or you for being gay?'

Will sighed. He'd been waiting for Charlie to ask that particular question. 'Um, not sure really. Ask Davey. He's better at that sort of thing than I am. Now, any more questions or can we start on the tomato seeds?'

'Oh yes. I love tomatoes.'

'I had noticed. Right then, fetch the packets. We'll use these seed trays for those.'

He reached for the special soil compost as Charlie picked up the three packets of seeds, and they started working on that summer's harvest. That evening, Will duly reported this conversation to David over a spaghetti dinner.

'So I've got to explain racial prejudice to him? Great. Thanks a lot.'

Will grinned and squeezed his hand. 'But you are so good at that sort of thing, honey.'

David shook his head in mock exasperation and went on, 'But you handled that well. The faggot thing, that is. I'm glad he asked though. Asked us, not one of his mates.'

'He trusts us.'

'He should, but most kids don't like talking to their parents about these things, or adults generally.'

'He talks to us about everything, doesn't he?'

'God yes. I still shudder thinking about the wanking conversation.'

During the Christmas school holidays the year before, he and Charlie had been playing chess in the living room. Lottie was upstairs doing the bathroom and Will was off buying wine and a new pair of jeans in Stroud.

Charlie had been looking very hesitant about something so David had said, 'Something on your mind, Charlie?'

'Mmmm.'

'What?'

'It's embarrassing.'

'Oh, I love that sort of conversation.'

He had winked and the boy had grinned, then exhaled in a mock, dramatic way. His chess opponent had said, 'Go on, ask me. I won't tell your gran.'

'Last night… well, I sort of wet the bed. Well, my pyjamas. But it wasn't wee.'

'Ah. You came then. That's sperm, from your balls. It's called an emission or ejaculation. It's totally natural.'

'But why did I have it?'

'Because you are maturing and developing your body for sex. We've all gone through that experience. Well, men have. It's all a part of growing up. It's called puberty.'

'But it was horrible.'

'Like that, yes, it is.'

'What do you mean, like that?'

It was now David's turn to go red. He hadn't realised he was going to have to have the sex chat so early. 'Um. Okay then. Right. You know about sex between a man and woman?'

'Yeah. The cock goes inside a woman and she has a baby.'

'Well, there's a bit more to it than that. You don't always get a baby, and the sex bit can be a lot of fun. The white fluid you got last night is your seed, and it can fertilise a woman's egg inside her womb, which is another part of her body, deep inside. But when you are young like you, your cock and balls can get excited all on their own, without a woman being there. And you get so excited, you end up with what happened last night.'

'But I was asleep.'

'Yes, but it can happen while you are sleeping. It's called a wet dream.'

'Ooh. Not sure I like that.'

'There are ways to stop it or reduce the number of times it happens.'

'How?'

Charlie looked at him so earnestly, David thought, "Oh well, in for a penny, in for a pound".

'You know your cock sometimes gets hard?'

'Yes.'

'Well, next time you are in bed and it gets hard, you can rub it up and down and make the fluid come out deliberately. It is called an orgasm. It can be very enjoyable. Some men really like it. You can think about a girl you consider very pretty while you are rubbing it. That can help. If your cock is dry, you can use a bit of spit, or some of your gran's Nivea cream
to make the rubbing smoother.'

'What do I rub it with?'

'Your hand. Um, like this.' He mimed the act and Charlie looked a bit startled.

'And that will stop it happening while I'm asleep?'

'It will cut it down, yes. Hopefully.'

'Cool. Why is it enjoyable?'

'You'll see. And use a tissue to wipe yourself afterwards, to keep the bed clean.'

'Okay. I'll try that. I don't want to wet myself in my sleep no more.'

A month later, Lottie had come into the study and said, 'Did you teach Charlie about wanking?'

'What?' gasped David.

'Did you teach my grandson how to have a wank? He says you did.'

He had gone red and explained the conversation. Lottie had snorted.

'Well, you could have warned me. He's going through jars of Nivea and all my Kleenex like a demon. And you should have told 'im to put the used tissues in the bin.'

'Where does he put them?'

'Under his bloody bed. It's like a cloud under there...'

David had burst out laughing and she had gone off, grumbling, 'It ain't funny if you're the one cleaning them up.'

He had bought Charlie a little wastepaper basket with a lid and told him the best way to tidy everything up without upsetting his grandmother.

Will laughed, 'Yeah, that was a triumph of communication, that was.'

'Not my finest hour, I'd admit...'

'Who taught you about wanking?'

'My father. He gave me a very medically sound version, as he was a doctor. And you?'

'My mate Ned.'

'The pig farmer?'

'Yep. In his dad's barn when we were twelve. He was a little bit older than the rest of us lads and he sat down on a hay bale, whipped his cock out and gave us a demonstration. There were five of us, standing there, watching.'

'Oh, kinky stuff.'

'Well, I did get a bit excited, but that were only cos of the look on his face. Maybe. He looked so happy when he came. I tried it that night and did what Charlie did. Got my pyjamas wet.

When mum found them, covered in dried spunk in the morning, she took the belt to me and told me it was the Devil's work.'
'It amazes me you turned out so sane and normal.'
'Me too. She was a nutter, weren't she?'
'Oh yes. No doubt about that, love.'
Charlie had gone to the grammar school in Stroud that autumn and, two years later, when he started dating a girl from Combe Weston, Will had sat him down in the greenhouse again, this time to talk about contraception. He had given the boy a packet of Durex condoms. Charlie had listened to the talk with a very serious look on his face, then started laughing.
'What's so funny?'
'I usually get mine by slipping round the back of the Fish and Trumpet and buying them from the vending machine in the gents. But thanks for these…'
Will's reply was unprintable.

Chapter Twenty-Three

On the first of June 1983, they celebrated their twenty-fifth anniversary. Although they had first met on the fifth of April 1958, it was the day they transitioned from employer and gardener to lovers that they remembered as their "special" date.

Paul Ford had organised a party in London earlier in May for their gay friends in town; his partner, Simon Higgins, who was the head pastry chef at the Savoy Hotel, made them a mock wedding cake with two figures of men on top. It was a very boozy affair, and most of the men there had just gone on and on about the fact they had lasted so long.

Back in Fletcher's Cross, they decided to arrange a more formal event, a dinner party. They asked Arty from the Frog to cook, Nell and a friend to serve, and invited Mary-Louise and Diana Bottomley, James Fallow and his wife, Murrey and Lillian Holt, and Lottie and Charlie. The seldom-used Wedgewood china was brought out and the silverware polished. For once, they bought some expensive wine, a nice 1979 Chablis from Robert Ampeau's vineyard to go with the starter, and six bottles of the very fine 1978 Chateau Neuf de Pape for the main course. The ten of them sat down to a wonderful meal of smoked salmon, roast duck and one of David's trifles.

The guests mixed well, chatting about all manner of topics until, just as the meal was coming to an end, Will asked Murrey how he had become David's agent. The old man chuckled and replied, 'Good question, Will. I was, and still am I must mention, a very successful agent when I received, in the post, David's first manuscript.

Now usually these were handled by one of my assistants, but on that day, I had a bad cold and left the office early. The postman was just coming in as I was leaving, it was raining and he was soaking wet, and he just thrust the package at me rather than go up the stairs to my office. I stuffed it into my briefcase and went home.'

He paused and drank some of the excellent wine, then went on, 'The next day I stayed home, but sat in my study and opened my briefcase to do some work, but discovered, in my daze the day before, that I had left the file I wanted to work on on my desk. All I had was this new book to look at. So, I started reading. It was wonderful. So exciting, so fresh, so detailed with a shocking ending that I absolutely loved. So, I wrote to "Connor Lord", care of David Manners, inviting him to lunch at Rules the following week. He wrote back accepting.'

David nodded and grinned. He knew what was coming next.

'There I was, sitting at my usual table, sipping a sherry, waiting for my new author. Because of the nature of the story, and the accuracy and detail, I was expecting an ex-military man, or perhaps a middle-aged civil servant, maybe even a spy doing a bit of ex-curricula writing on the side. What I got was a bloody schoolboy!'

The vicar gasped, and the rest of Murrey's audience laughed.

Charlie nudged Will and whispered, 'He said bloody.'

'This tall, handsome young man in a blue blazer and Oxford bags walks up to my table and says, "Good afternoon, Mr. Holt. I'm Connor Lord." He was eighteen. Eighteen! I couldn't believe it, but we shook hands, he sat down, asked for a glass of cider and started talking about the book and his target readership.

I was mesmerized. He was so adult and so confident. I asked him how the hell he had got so much detail about the war and its aftermath, and he said, "I'm good at listening to people and asking questions. There were a lot of ex-military officers and servicemen who returned to Salisbury after the war and I talked to them. I talked to the spivs selling black market rations in the town and put it all together after that. I started asking questions and listening when I was fifteen and *The London Connection* is the result.'

'God, I must have seemed so arrogant, Murrey.'

'No, David, not a bit, just confident. I liked that, I loved the book, and I knew it would sell.

If you remember, I then said I knew of two publishers who might take it, Hilton and The Norfolk Press. That was when you asked me what my commission would be.'

'Oh yes…'

Murrey looked around the table and said, 'I told him I would take twenty per cent of his royalties. Pretty standard at the time. He just looked at me across the table… we had started lunch at this point and were eating a damn good chicken stew. There was rationing back then and it was all they had. He looked at me and said, "No, ten per cent." I gasped and said that I wouldn't get out of bed for ten per cent. He just smiled and said, "I don't care where you do the work, Mr. Holt, but you're only getting ten per cent.".'

Everyone around the table laughed.

Diana Bottomley asked, 'Oh, Murrey. What did you say?'

Holt shook his head. 'Well, to be perfectly honest, Diana, I was so damn impressed by that line that I roared with laughter and told him yes. I could tell he was going to make me a lot of money, even at ten per cent. He was a natural talent. And I was right.'

David added, 'He was. I've made him stinking rich. He's driving his third Rolls Royce paid for by my royalties!'

Lottie asked, 'I'm not exactly sure what an agent does, Murrey. If David's publishers like his books, why does he need you?'

'Ha. An excellent question, dear lady. Firstly, many publishers don't like dealing with authors directly, in case they have bad news for them. They prefer people like me to give it to the little darlings. Secondly, there are different contracts to negotiate. The American rights, translation contracts, other overseas publishing rights in different countries and languages. Film and television rights. There is an awful lot of legal work to be done to make sure David's work is protected, paid for and not copied.'

David stepped in to support him.

'Murrey saves me from having to deal with any of that, Lottie. He's brilliant at the legal and contract side of things. He leaves me completely free to do what I love most, which is write. That's worth ten per cent of anyone's money.'

There were smiles and murmurs of appreciation. Nell and her friend came in and cleared the dessert plates, brought coffee and decanters of port and brandy, a box of "After Eight" mints and left them alone again. Will tapped on the side of his brandy glass and said, 'Um, you all know why we invited you here tonight.

Now normally Davey is the one who does all the talkin' but I thought I would say a few words for a change.'

His lover grinned. It was true. He tended to be the one with the words.

Will went on, 'Twenty-five years ago, I was a bit of a sad case. I was closeted. I knew what I was and what I wanted, but never thought I would find it in Fletcher's Cross. Then Davey arrived, I got drunk and here we are, still together after all these years. He's the best thing that ever happened to me… and, apparently, the best thing that ever happened to you too, Murrey.'

The old agent nodded, 'Damn right, old boy.'

'We wanted to share this moment with the people who have made our lives even better. Lottie. You have looked after us so well for all that time and we thank you so much for that, and your friendship as well. Charlie. You're the son we never had. Diana and Mary-Louise. Your friendship means so much to us. And James and Mary. Through your church and your warmth, we have felt part of this wonderful community. You have never judged us or rejected us and we thank you for that too.'

The vicar and his wife looked slightly embarrassed but nodded. Will continued, 'David. I love you. I can't say it in church yet, but I do.'

David got up, came around the table and hugged him, whispering, 'And I love you too, Will.'

Their guests clapped as Mary-Louise said, 'Hear, hear.'

James said, 'That was lovely, Will. And, one day, I hope you can say it in church.'

'Thank you for that, James.'

At eleven the taxi arrived to drive Mary-Louise and Diana back to Swindon. Lillian, who never drank, drove her husband home, and Lottie and Charlie walked back to their cottage. The chef and his team had already left, well tipped. James and Mary Fallow were the last to leave. They stood at the front door shaking hands, and James said, 'I just wanted to say that, if anyone ever doubts that gay men can't have long, loving relationships, they should meet you two. And I meant what I said. I do hope you can get married in church one day, preferably mine!'

They went off and David closed the door and took Will in his arms.

'Very nice speech.'

'You approved then?'

'I have always approved of you and everything you have ever done or said.'

They kissed.

David added, 'Time for bed?'

'After we've let the dog out.'

'Of course. Mustn't forget Henry…'

Chapter Twenty-Four

On the seventeenth of December that same year, an event occurred that could have ended this story of the happy couple very abruptly. David was nearly killed in a terrorist attack in central London. Since the mid-seventies, Britain had experienced several terrible atrocities as a result of the Provisional Irish Republican Army's, or IRA, campaign to get the British out of Northern Ireland and form a united Ireland. London had been the scene of many of these including the Chelsea Barracks bombing in 1981, and the Hyde Park and Regent's Park bombings in 1982, that killed eleven soldiers and five horses.

None were as bad as the Birmingham pub bombings in the seventies, that killed twenty-one people, but many lived in fear of a bomb in a waste bin, a parked car or on a train. They had already carried out one attack that month in the capital, on the tenth of December, at the Royal Artillery Barracks, that had injured three soldiers.

David and Will had to go up to London for the Hilton Publishing Christmas party on the sixteenth, a Friday, and they had discussed the possible threats but not taken them too seriously.

'If we don't go, they will have won,' was the author's response. However, Will woke up with a bad cold on the Friday morning, so David went up alone, staying at Brown's Hotel as usual. On Saturday morning,

after breakfast, he went shopping for some last-minute items for the holiday season, visiting the HMV shop to buy a copy of the 12-inch single of Frankie goes to Hollywood's *Relax*, Fortnum and Mason's for some Royal Blend tea, then he headed into Harrod's to get Charlie new t-shirts, and some Roka cheese biscuits and a round of Stilton from the Food Hall. He arrived at twelve thirty-eight in the afternoon.

At twelve forty-four, a man using an IRA codeword phoned the central London branch of the Samaritans charity. The caller said there was a car bomb outside Harrods and another bomb inside the department store and gave the startled charity worker the car's registration number. According to police, he did not give any other description of the car, or mention in which of the four streets around the department store it was parked.

Harrod's on the Saturday before Christmas was packed solid with shoppers. There were queues at every counter, and most visitors were laden with shopping bags filled with their Christmas shopping, party frocks, toys, and other gifts. Many had small dogs on leads, as the store welcomed the pets of patrons who would never consider leaving them at home. In the Toy Department, there was a very long queue of parents with children waiting to speak to Santa Claus in his grotto.

David was standing waiting to be served in the Food Hall on the ground floor at ten past one when the first announcement to evacuate the building came over the Tanoy system. There had been delays in the warning reaching the store. The noise level was such that many people missed it and, when it was repeated, it took time for them to register the fact they were being asked to leave. Some just stood still, not wanting to lose their positions in the queues, while others started to panic and rush to the exits.

David heard the announcement the second time it was given and started moving towards the exit, but the sheer number of visitors hampered his progress. He had just reached the entrance of the designer menswear department when the bomb exploded at twenty-one minutes past one, as four police officers in a car, an officer on foot and a police dog-handler neared the suspect vehicle, which was parked directly outside the store, on Hans Crescent. Just outside the menswear department.

The effect of the bomb inside the store was dramatic. The big windows facing the street blew in and demolished the Christmas displays, sending debris and shards of glass over the tightly packed shoppers trying to escape. Ceiling tiles and some light fittings fell on top of the people, who had been blown off their feet by the shock wave, and clothes racks with shirts and jackets added to the chaos. There was dust everywhere, and it was hard to see clearly. David, who had been standing in the doorway from the Food Hall side, was blown over, his ears ringing from the blast.

He fell on top of a large woman, who had a small, white poodle on a lead with her. The dog, terrified by the blast, bit David's wrist as he struggled to his feet and tried to help the poodle's owner up. The window displays had protected most people from the flying glass, and now they tried to help their fellow Christmas shoppers who had been injured. David, still holding his bags, got several women to their feet and led the way to another exit, on the opposite side of the store. People were suddenly very quiet.

The store's staff, who had been given training in evacuation procedures, were very professional, and were then aided when more help arrived, police officers, firemen and some ambulance crews.

If the public had known the warning had been for two bombs, one inside the store, they might not have been so calm, but the police and firemen were aware of this additional danger, so they acted quickly to get everyone out. There were many children amongst the flood of people leaving, most crying in shock, but some just miserable that they had missed out on giving Father Christmas their wish lists.

A second bomb did not detonate.

Outside, police officers, themselves still devastated at the loss of colleagues, shepherded people away from Harrod's. David looked around for a telephone box but couldn't see one. He managed to hail a taxi and went back to Brown's Hotel, because he needed to collect his luggage.

The manager gave him first aid for the dog bite, then called the hotel's doctor who came and gave the injured man a tetanus shot, "just in case".

David borrowed the manager's phone to call home. It was two fifty before he could finally get through to Fletcher's Cross.

Will had spent most of that day on the sofa, watching television. He had surrounded himself with essential items, like a big bottle of Lucozade, a pork pie, a box of Kleenex tissues, aspirins and the handset of their latest gadget, an AT&T cordless telephone. He was still there, watching *It's a wonderful life* on BBC2, when Lottie called at one-forty. She had been listening to the radio whilst knitting some socks and heard about the Harrod's bomb.

'Is David back yet?'

'No. Why?'

'There's been another bomb go off in London. Outside Harrod's'

'Oh God, no. Are there many injuries?'

'They ain't saying at the moment, but some policemen were killed. They think.'

'Oh God.'

'Was David going to Harrod's?'

'Yes. He wanted some of their stilton.'

'Oh dear. Well, I hope he's alright.'

'He might be trying to call. Can you hang up and I'll let you know if I hear anything, Lottie?'

'Of course, dear. I hope he's alright.'

Will lay back on the sofa in a total panic. He had never even considered what his life would be like without David, and he now started crying uncontrollably at the thought he might spend the rest of his life without him. Henry woke up and stuck his nose under the blanket, and Will dragged the dog onto the sofa and hugged him. Over an hour passed before the phone rang again and a very welcome voice said, 'I'm alive. In case you were worried.'

'Of course, I've been worried. Are you okay? Were you there? In Harrod's?'

'Yes. Right inside. Look, I'm going to take a taxi to Paddington now and catch the three-fifty train. I'll tell you all about it when I get home. How's the cold?'

'Awful, but I'm feeling much better now I know you haven't been blown to bits.'

David chuckled down the phone and said, 'I'll pick up an Indian take-away on my way from Stroud. A hot curry will do you good. Bye, love.'

'Bye, Davey. I'm so glad you're okay.'

Will immediately called Lottie to tell her the good news.

David managed to find the last free seat in the carriage. He had booked for an earlier train so was happy to get any seat at all. Many people were standing as the train left the station. He was sitting opposite a man who was studying him carefully. He realized he must look a bit of a mess. Dusty from the blast, with a bandage on his wrist, his shirt grubby too.

'You look as if you've been in the wars,' said the man.

'I have. I was in Harrod's.'

The man nodded then showed him the palm of his right hand. There was a white mark, right in the centre and, when he turned it over, a corresponding mark on the top.

'I was on the road that runs by Hyde Park when those bastards blew up a car and killed those soldiers and horses. They used a nail bomb, the fuckers. Got one right through my hand. I'd stopped to watch the Household Cavalry parading by when all hell broke loose. Still got a ringing in my ears some days.'

'You were very lucky. There were a lot of terrible injuries that day. So was I today, I think. It was chaos in the store after the bomb went off. Dust everywhere, and children, adults and bloody dogs all over the place. They say six policemen were killed.'

The man sighed heavily. 'What bastards they are. At Christmas too, with all those kiddies around.'

They discussed the Troubles in Northern Ireland and the difficulties of resolving the conflict until the man disembarked at Swindon. David got into his Range Rover at Stroud station, stopped off at the Indian Star restaurant for food and arrived back home just before seven thirty.

He dumped his luggage and packages on kitchen floor, put the food in the oven and went into the living room. Will burst into tears as soon as he saw him and they hugged for fifteen minutes before he stopped crying.

'I thought I'd lost you.'

'I'm okay. Really, I am.'

'I've never even considered what I'd do without you.'

'Okay, well, I'm back. You don't have to think about that sort of thing any more. I'm fine.'

Will looked at his bandaged wrist.

'You've been hurt.'

'That was a dog, not the bomb.'

'A dog? Are they using dogs now?'

'No, you twat. Some woman's hysterical poodle bit me in the men's department.'

'Where?'

'In the menswear department.'

Will grinned, his sense of humour returning.

'I thought you meant it had bitten you down there...'

David shook his head in mock-disgust, 'Hmmm, you're obviously feeling better. Hungry?'

'Actually I am.'

The morning newspapers all had pictures of the damage to London's most famous department store on the front pages. It looked devastated. All five floors on the Hans Crescent side had had their windows blown out, and over one hundred people had been treated in hospital for injuries from flying glass, metal fragments and other debris from the blast. The remains of the car were scattered all over the street, and the other vehicles that had been damaged by the blast.

There were pictures too of the six police officers killed in the line of duty. Ordinary men, going about their duty, trying to protect the citizens of London.

'What utter shits those people are, to do something like this just before Christmas,' said Will.

The bomb was later discovered to have been planted by a splinter group and was not "authorised"; the official IRA issued a statement condemning the bombing as it might have killed children, but most people thought that was simply a PR exercise.

Will and David didn't go to London once during 1984, not wishing to tempt fate. But eventually they returned, like everyone else, determined not to let the terrorists win.

When the IRA bombed the Grand Brighton Hotel during the Conservative Party annual conference on the night of the twelfth of October that year, trying to assassinate Mrs Thatcher and her government, it was the only time David felt sorry for her.

For the first time, they had something in common and, for once, her experience was far worse than his.

1990

Chapter Twenty-Five

Chelsea Flower Show, London, 27th May

'Would it be possible to get your autograph, Will?' said the plump woman in a dress covered in a bright red clematis print.

'Of course. Who shall I write it to?'

'Lorna.'

'There you go, Lorna.'

'Oh, thank you. I love your programme. And your column.'

He had signed five autographs since they had had lunch in the Royal Horticultural Society's President's Pavilion at the main flower show in the English calendar. It was being held in the grounds of the Royal Hospital Chelsea as usual, the home of the famous Chelsea Pensioners. David was getting used to his lover's fans swarming around them at such events. As they moved towards the exhibition of special gardens, Will said, 'You're very patient.'

'I'm glad you are a success and popular. It's wonderful.'

'You really mean that, don't you?'

'I do. You had to put up with being Connor Lord's boyfriend for years. It's nice you get main billing. I love being Will Forman's partner.'

Will was very popular with *Gardeners' World* viewers for a variety of reasons. He never talked down to people or dumbed down either. His gentle, West Country dialect, his twinkling blue eyes and that fact he still looked much younger than his fifty-five years made him a figure of adoration amongst middle-aged housewives, and quite a lot of gay men too.

Will got stopped by another woman who wanted him to sign the Royal Horticultural Society's show catalogue for "her daughter", before they walked through the section he had visited the day before in his capacity as a judge. The results would be announced later. A very slim man in a shocking orange shirt waved to them and said, 'There you boys are. I've been waiting hours.'

'Don't exaggerate, Sid. You arrived after us. We saw you.'

'Oh. Bitch. Well, here it is, David. The Lighthouse Garden.'

They wandered around the stunning display, produced by the helpers of those living at the Lighthouse Hospice, for gay men suffering, and dying, from AIDS. David had given thousands of pounds to support their work, and Will had suggested this garden as a tribute to another famous garden expert, who had died from the terrible illness whilst staying at the Lighthouse. The rainbow colours had been incorporated beautifully into the design, picked out by the blooms chosen. David said, 'It's wonderful, Sid. A really beautiful piece of work.'

'Sadly, Morty and Keith aren't here to see it. They produced the original design. Do you think it will win?'

Will sighed and said, 'Not sure. You have a lot of competition this year. If I was judging purely on spirit, yes, but you're up against Boy Scouts this year.'

'Not for the first time...'

David rolled his eyes and muttered, 'Careful...'

Sid grinned, 'The old ones are the best.'

'Still talking about boy scouts?'

Their host suddenly waved to a figure in the crowd of visitors. 'Oh, there's Martin. Must have a word...'

He ran off, and they carried on their tour of the other gardens. After more stops for autographs, they reached the tea tent and managed to find a table for two in the crowded space. David exhaled deeply, 'God, it's hot. How much longer do we have to stay?'

'We can head home if you want. I've finished all my stuff, and I don't have to be here for the presentations. Angela Rippon's doing those.'

Will was now a regular presenter on *Gardeners' World*, and was very popular, focusing on the environmental importance of planting in the garden. He was also a regular guest on the BBC Radio 4 show, *Gardeners' Question Time*, whenever it came to the West Country.

He had been given honoury membership of the RHS, and this was his first year as a judge. But he had found all the attention a bit overwhelming and had had enough for that week.

'Okay, quick cup of tea then we head for the hotel to get our stuff. We can be on the four thirty easily.'

'Home by eight. Lovely,' Will replied.

The waitress took their order and came back five minutes later with a pot of tea for two, two scones, some clotted cream and a little saucer of strawberry jam. The price made David wince. But that was the Chelsea Flower Show for you.

As they were eating, a woman came up, clutching a copy of *Forman's Flowers for Bees and Butterflies*, his third book on garden plants. Written by David. She hesitated then said, in a hushed voice, 'Oh, I am so sorry to interrupt your tea, Will, but I missed your book signing this morning. Would you…?'

She handed him the book and a biro. He said, 'No problem. Hope you like it. Who should I write the inscription to?'

'Sonia.'

He wrote a few words, signed his name and handed it back. She read his note, smiled, said, 'Oh, thank you so much,' and rushed away. Will turned to David and said, 'You should have signed that. You wrote it.'

'True, but I only put down your words.'

'I still think it's cheating.'

'Bollocks. And I don't mind at all. It's a pleasure.'

Will squeezed his knee under the table and finished his scone. He had signed over four hundred copies between nine and twelve that morning. He was being published by Gloria Hilton, of course. All three of "his" books had been best-sellers, and he had a very healthy bank account as a result. He was also the public face of a potting compost company and a seed producer, which added to his growing wealth. But he, like David, had made big contributions to the AIDS charities around London. They always needed more money.

The train back to Stroud was packed when it pulled out of Paddington but eased after each stop as the daily commuters disembarked. They were left alone for the most part, although a group of women who had been at the flower show noticed Will and asked for autographs, then interrogated him about the best flowers for encouraging more insects into their gardens.

They now had five beehives in the orchard at Wisteria House and ate their own honey regularly. Cal had added beekeeping to the growing list of his skills.

The eighties had been a mixed decade for the couple. David had published *Smoking Gun* at the end of 1982, and it had remained on or near the top of the New York Times best sellers list for two years. It had led to enquiries, more lawsuits (not directed at him thankfully), tougher safety warnings on cigarette packets but no real reduction in sales. Smoking might have decreased in Europe and North America, but the tobacco companies had simply focused on Asia, Africa and Latin America instead, as David predicted in his book that they would.

At the same time, Will's star had been in the ascendant. People loved his little slots on *Gardeners' World*, and he had become one of the regular presenters. He was away a couple of days each week during the summer, filming around the country, leaving Cal to take care of the garden.

After *Smoking Gun*, David had taken a break, apart from ghost writing Will's flower books. He wasn't suffering from writer's block, but just didn't feel excited by any topic. Henry the first had died of old age, and he trained up Henry the second, another golden retriever, answered his ever-growing correspondence and wrote thoughtful letters to the Times and the Guardian about environmental matters. Ginny still worked for him two days a week.

When the AIDS crisis hit hard, between 1984 and 1987, the public's mood towards gay people swung all over the place; whilst some blamed the men themselves for getting infected due to their, according to the Daily Mail, rampant sexual activity, others joined together to offer support and assistance. When the London Lighthouse opened in 1986, David was a major financial donor, as his editor, Paul Ford, was one of the first people in England to contract, and die of, an AIDS related illness.

They missed Paul very much as a friend and David had greatly valued his contributions to his novels since the very beginning. When a well-known newspaper columnist wrote a vile opinion piece saying public funds should not be used to treat AIDS victims, as they were abominations in the eyes of God, and had caused their sickness by their lifestyle choices, David wrote a now famous response to the Times.

In it he said that, if such an idea was logically extended, and people should be refused medical care due to their lifestyles, the National Health Service should stop treating sports injuries, car crash victims, cyclists who had fallen off their bikes, smokers, obese people, drunks, pregnant women… in fact, he wrote, they should only treat healthy people. It made fun of the columnist's views in a lethal way, and his newspaper column, and his career, ended abruptly. He moved to America and started working for a talk radio station in Texas, ranting into the night about gays, immigrants and commies.

Chapter Twenty-Six

Britain in the eighties was a country of contrasts. There was extreme poverty alongside a surge in wealth brought about by a property boom; on the streets conspicuous consumption was the name of the game. Hugo Boss suits were the dress de jour for men, the very glamorous Stringfellow's nightclub opened in London, and Japanese designers, Next and other fashion brands filled the high streets. But unemployment was rising and homelessness was growing, and the numbers of young people sleeping rough on the streets of London a scandal. David and Will both gave money to Centrepoint, a charity for helping them.

Margaret Thatcher was at the height of her power. Due to some reactionary members of her inner circle, shocked by the gay AIDS "plague" as pushed by their favourite right-wing media she, in her typical, arrogant, self-righteous assertive style, brought in Clause 28, banning certain books being used in the classroom that showed gay people in a positive light, and any constructive discussions about gay people being taught in schools.

David, or rather Connor Lord, was a vocal opponent and was invited to the Oxford Union to talk about it, and other matters. He reminded his audience that Margaret Thatcher had voted in favour of the '67 Act and said, "Perhaps the Prime Minister would like to go back to the days we when only existed at home, behind locked doors, with the curtains closed."

He spoke of the injustice and danger of preventing schools from teaching any subject that was of importance to children in preparing them for the world, and also discussed the need for more work to prevent an environmental disaster.

The room was packed to the rafters for his speech, and he answered questions for two hours afterwards. To say he was a success was an understatement.

In 1987 he had finally started writing again, a political thriller called *Here is the News*, about a right-wing religious conspiracy in the near future. The story described hidden groups trying to influence the development of society through advertising, branding, buying TV and radio channels, and the media generally. He came up with two startling ideas.

One was that every person had become a reporter due to them having cameras in their mobile phones. As that feature didn't exist at the time, several critics said the idea was a bit too far-fetched. He also came up with even more outrageous idea that opinion could be presented as real news, thus influencing the way people thought, through news channels that told people that whatever they believed, however stupid or outrageous, was real.

Published at the end of 1988, it went straight to the top of the best seller lists, proving he still had it when it came to writing books people wanted to read.

One of the major film companies in America made a bid for the film rights to *Here is the News*, but Murrey Holt got a message from a company whistle-blower that the only reason they were bidding was so they could block the film from being made. The company was now part of a major global empire with a very dominant Australian owner. The offer was rejected and eventually HBO would turn it into a mini-series in 1997.

And it would be another three years after that before Sharp, the Japanese electronics corporation, produced the first mobile phone with a camera in it...

David was now nearly sixty. With a cropped silver-grey head of hair and good bone structure, he still looked younger than his years. He had stayed fit, by daily exercise and dog walking, seldom drank and ate good, organic, healthy food. So did Will. He was equally well-preserved, and still had a full head of dark blond hair.

They had remained faithful throughout the period, not because they were afraid of becoming infected with HIV, but it simply never occurred to them to go and find anyone else to have sex with. The closest they had come had been in San Francisco in 1988, when they had gone over for the Gay Pride Festival.

A very young and cute gay author and poet called Jules Daniels had recognised David and approached them at a bar overlooking the main drag on Castro. He had chatted them both up expertly. He said he loved "daddies" and would love to serve them if they were interested. He looked like one of the blond surfers they had watched in those early gay porn videos. He was a real delight to look at and talk to, flirted with them blatantly, and they came very close to breaking their duck as far as a threesome was concerned. At the very last moment, Jules had said, 'And you don't have to use condoms, you can bareback me to your heart's content.'

It was like the coldest bucket of water being thrown over them. David had looked at the boy in disbelief and asked, 'Do you allow men to have unprotected sex with you?'

'All the time. I'm a bug chaser. If I get it, I get it. Much prefer sex without rubber. Don't you?'

'Only because we have never had sex with anyone else. If we did, we would always use a condom. It is utter madness not to.'

'Ah, that's crap, dude. Your loss.'

The boy had turned and flounced away and the two Englishmen had finished their drinks and walked slowly back to their hotel, mentally shaking their heads at the stupidity.

So many gay men had died in the preceding five years, and yet that lad was prepared to get infected, and spread it too. Most men they met during that trip were very focused on staying healthy, and those who had been a bit rampant before the crisis had changed their ways and only had safe sex, relieved they had dodged a bullet. They had lost friends in England and knew of many that had died in America as well, including some from their visit to the Big Apple.

David and Will had heard that Alex, their New York City guide, had died in 1986. The famous actor and the TV evangelist had both passed away in 1987. The Republican senator had committed suicide that same year, and his autopsy showed he had been infected as well. Finally, five months before the Pride event in California, Tony Fleck had died of an AIDS-related illness at his palatial summer home in Nice, France.

One of the worst aspects was that many of those who died were closeted before, so their families had to cope with their deaths and a very public coming-out process at the same time. It was a heart-breaking time for gay communities and their relatives on both sides of the Atlantic. Sadly, because the focus had been on AIDS being a problem limited to homosexuals in many people's eyes, the number of straight men and women dying rose as well, as they thought the infection was limited to the homosexual community.

Around the world, health services were struggling to inform people of the risks, but met prejudice and hate, or simple, down-right ignorance.

David and Will's support for the Lighthouse was not their only contribution. David wrote several opinion pieces, for the Times and the Guardian, emphasising the need for better information for straight men and women, couples, and younger gay men. His voice was one of many, but the numbers still went up, until finally drugs were available to combat, but not cure, the disease. Ironically, those drugs managed to keep people alive as long as they took them every day for the rest of their lives. The ideal medical product for a drug company…

*

They lost two other old friends in 1989, from other causes. Sir Peter Scott died of heart failure aged eighty, and Diana Bottomley passed away too.

She caught a bad case of flu, which turned into double pneumonia and, despite receiving the best possible care from Mary-Louise, she died in March. They went to her funeral in a tiny church in a village outside Swindon, with David giving one of the readings. The service was well attended, by women; they were the only two men there, apart from the vicar. They kept in touch with Mary-Louise for five years, before she died of breast cancer.

Chapter Twenty-Seven

Will was away in Devon and Cornwall the week after the Chelsea Flower Show, filming. David and Ginny were very relaxed for a change. The mail generated by *Here is the News* was beginning to die down, so they were able to focus on his archive material once more. He enjoyed drifting back through time as he read old letters from Ian Fleming, other authors, old newspaper articles, and some of his past reviews. The banana boxes of old had been replaced by box files, all neatly labelled, so it was an easy task to find things now, thanks to his wonderful secretary.

On the Wednesday, Lottie knocked on the study door and asked for a word. She sat down and sighed heavily.

'I've got to stop working, love.'

'About bloody time. I've been waiting for you to decide that. You don't need to work. You've got a great pension.'

'I know. You've been very generous.'

'Bollocks. You've looked after us so damn well for over thirty years. It's the least we could do.'

She was now seventy-one and should have stopped five years before at least. He had raised the subject at the time, but she had glared at him and said, "Nothing wrong with my legs or my arms, and I'd just get bored." Now she did look tired and he asked,

'Is everything all right? You're not ill?'

'No. Just exhausted, that's all.'

'When did you last see a doctor?'

'1977.'

'Do you think perhaps you should?'

She shook her head. 'No. I'm fine. If I feel ill I will, don't worry, but we need to find you a new cleaner. Unless Her Ladyship is going to get the hoover out.'

Ginny snorted and said, 'In your dreams, you old bag.'

They had developed a love hate relationship over the years and enjoyed the banter. Now Ginny added, 'We should ask Nell. She might know someone.'

Cal and Nell, now married, lived in one of the cottages next door, with their two girls, aged five and four. Nell still worked three nights a week at the Frog but had wanted a job with a bit more social conscience and did two days a week at the Citizens Advice in Studley Combe. Like Lottie, she didn't need to work, as Cal was well paid, but she got very bored once the girls were at the nursery school.

'That's a good idea. Surprising coming from you...' said Lottie, then winked at her friend. Ginny's reply was very short, and rude. Then Lottie looked at the drawings on the desk and asked, 'So, this is it, is it? We're all set then?'

'Yep. We fly out on the 15th of June at nine in the morning from Heathrow. The builders will arrive then and have it finished by the 22nd. We get back on the 24th.'

The drawings were plans and elevations for a huge conservatory to be built where the terrace stood. That would be extended into the lawn, and the two-storey white structure attached to the back of the house, centred where the living room doors led out into the garden. It was David's

birthday present to Will, and the gardener, who had wanted such a conservatory for years, knew nothing about it. They were going on holiday to Norway, their first ever visit, starting in Oslo, then flying up to a town in the Arctic circle, and driving back down to the Norwegian capital again while it was going to be put up.

Lottie tutted. 'You think they can do all that in seven days?'

'That's what they said. They are specialists in building these things, nothing else. They are experts.'

'I've heard that before.'

'They promised to clean up afterwards too.'

'That I've heard before as well...'

She was referring to the renovations done to Wisteria House in 1988. The first major work done since David had moved in back in 1958.

All the windows had been replaced and now had triple glazing to keep the heat in. A new roof had been put on, also well insulated. Both bathrooms had been updated, including a big walk-in shower in the one attached to the master bedroom. The kitchen had been ripped out, with a beautiful Smallbone replacement installed in its place. The old Aga had gone; they now had a six-ring gas stove and an eye level oven. The old scullery was renovated too and boasted the latest front-loading washing machine and a tumble dryer, which had made Lottie's life much easier. The work had taken much longer than planned, and driven Lottie mad, hence her comment about the new conservatory.

David looked at his oldest friend in Fletcher's Cross and said, 'They'll be fine. Just make sure the builders don't crush Albert. He does tend to get underfoot with strangers.'

Albert was the latest member of the Wisteria House family. In March, David had bought some duck eggs from the farm shop, leaving them in a large mixing bowl on the kitchen worktop overnight, planning to use them for a cake recipe the following day. They had been woken up at five in the morning by Henry the Second barking like mad and gone down to find one of the eggs had hatched. They showed the duckling to the dog and they instantly bonded. From then on, the duck would sleep in Henry's bed and follow him around everywhere.

They managed to feed him and keep him alive, then got him swimming in the pond. He grew into a beautiful white Aylesbury duck, who had free range in the house and the garden. He was exceptionally tame and would come into the kitchen all the time when Lottie was there, as he thought she was lovely.

Lottie rolled her eyes and replied, 'That ruddy duck...'

The author looked out of the window down the garden. Albert was swimming on the pond, with Henry the Second keeping guard, crouched by the water's edge, watching his friend have a nice paddle.

*

Nell came up trumps and found them a twenty-two-year-old woman, living in Studley Combe, who needed a cleaning job. She had a son and, like Flora had been, was a single mother, not because of an orgy but the result of a one-night stand.

Her name was Nicky Stanley, and she worked alongside Lottie for a week to get to know the house and her new employers' demands before they went off on holiday. Nicky's parents had moved to England from the West Indies in the fifties, and her dad was a bus driver along the Stroud - Studley Combe route. When the council houses were being built, he had applied for one and still lived there with his wife, daughter and grandson.

David was a little worried about them having a black cleaner, not because he was prejudiced but he wasn't sure about the optics. Nicky made it clear she loved cleaning and didn't think the less of them. Far from it.

'I need a job, and I'll be doing something I love doing. You like writing, Will likes gardening; I love cleaning.'

They went off to Norway on the fifteenth of June, and the three women of Wisteria House kept a very close eye on the builders when they arrived to move the terrace and put up the conservatory. Lottie, in particular, was worried about the artwork being stolen and made sure the house was locked up tight each night. Henry the Second stayed with Cal and Nell, as he was too boisterous for her to manage.

David and Will flew into Oslo and had one, incredibly expensive, night in that city before flying up to Narvik in the far north of the country, where they rented a car. They drove slowly down through Norway, along the coast road, looking at the fjords and stopping once to go whale watching.

They drove slowly because the road was only two lanes, and ran through many small villages and towns, with speed limits of 20, 30 and 40 kilometres an hour. They were constantly speeding up and slowing down, and if they got stuck behind a caravan, they stayed stuck until it turned off at one of the many camping sites along the route. It was very frustrating, but they enjoyed the views. They had visited Sweden and Denmark, but this was their first time in Norway.

Their choice of Norway was partly due to their love of *The Hitchhiker's Guide to the Galaxy*. In the series, highly intelligent aliens (in the form of two white mice) had built a supercomputer to try and find out the question that Deep Thought's answer of 42 related to. This computer was the Earth, and Slartibartfast, one of the characters, was the individual who had designed the "crinkly bits" down the Norwegian coast.

It was a stupid reason for going somewhere on holiday, but they couldn't care less.

They finished with another night in Oslo, visiting the famous London gay bar after a dinner of a huge bowl of prawns, brie cheese and French bread on the waterfront.

The London was packed with a real mixture of young and old, sexy and not so sexy, men, and they were made to feel welcome. Not as a famous couple, but simply as two Englishmen on holiday. Once again, they didn't take anyone back to their hotel after the drinks, although a couple made it clear they were very available should they want to.

They arrived back in Fletcher's Cross at four in the afternoon the next day. As David parked in front of the garage, Will glanced to his left and gasped, 'What the hell is that?'

The huge conservatory had been finished as promised. Its Georgian style matched the house perfectly and the builders had done a great job in not damaging the lawn. The old terrace had been lengthened and some new flagstones found that matched to fill in the edges around the new, white-timbered structure. It looked as if it had always been there.

'That, my love, is your birthday present.'

'Really? Oh, Davey...'

He leapt out of the Range Rover, dashed through the back door, ignored the three women having tea in the kitchen and ran through to the living room. The French windows stood open and he stepped into his gift.

The floor was made up of honey-coloured stone tiles. In the centre stood a large round white wrought iron table, and four white chairs made of the same material with deep navy-blue cushions. On either side stood a huge stone pot with a mature lemon tree in it, and on each side of the doors leading to the garden, another, smaller stone pot held a bay tree, clipped to resemble a round lollipop. Will just stood there, mouth open, staring happily around him. David came up behind and hugged him.

'Like it?'

'It's perfect. It's bloody perfect. Oh, thank you, Davey. Thank you so much. I can fit loads more plants in here.'

'I know. I bought these four just to get you started.'

'Where are they from?'

'Hillier's, near Winchester.'

That was one of the best growers in the country. They had supplied the gingkoes by the pond, which now stood nearly twenty feet high. Will nodded and said, 'Well, they must be good then. Oh wow.'

He hugged him and gave him a smacking great kiss. Nicky came in, carrying a tray of tea and placed it on the table.

'Sit your arses down and have a cuppa. You must be knackered after your flight home. It's nice, isn't it, Will?'

'Nice? It's fabulous.'

David, down on one knee cuddling Henry and Albert, both ecstatic their masters were home again, asked, 'Were there any problems, Lottie?'

Ginny and their old friend had joined them.

'No, none at all. They were really professional. And they worked so neatly and fast. It's lovely, isn't it?'

'It is.'

Will turned and asked, 'How did you plan all this? I didn't have a clue.'

David laughed. 'Well, it helps that you're a TV star now and are away sometimes. I got planning permission, including an inspection by the county architect, the design done and the work sorted out while you were off filming. And they promised they could do it in seven days, hence the holiday.'

'And you lot didn't say anything...' Will said.

Lottie smiled, 'The one thing we are all good at here is keeping secrets, love.'

They held Will's birthday party the next day. Friends from *Gardeners' World*, *Country Life* magazine, the Lighthouse and the village all came.

David had booked and paid for rooms at Langham House Hotel (Huxton had moved on and the country mansion had become a fine hotel) for some of the guests.

Wisteria House was filled with laughter and noise, and the conservatory was the star of the show. After midnight, when the last guests had been piled into taxis, the two men stood in the new space, arms around each other, just hugging.

'That's the best party I've ever had, and the best present too.'

'Oh, I thought your boots were the best.'

'They were then, but you've outdone yourself this time.'

'You're worth it.'

'How can I thank you?'

David whispered a suggestion into his ear and Will chuckled. 'Oh, that's easy. You might even get two of those...'

Chapter Twenty-Eight

During the 1980s, the gardens around Wisteria House had also become well-known. Even if they were not open to the public, many of Will's contributions to *Gardeners' World* were filmed there. He would stand in the greenhouse and talk, or work in the vegetable garden, the orchard, or around his compost heaps, sharing his expertise with fellow garden lovers.

During these recordings, David stayed out of sight. Even if the production crew knew he and Will were lovers, the producers of the programme didn't want to push that side of their star, especially during the AIDS crisis. Even if Connor Lord was out, most people didn't connect him to David Manners, or Will for that matter. David didn't mind the house and grounds being used for the programme, but he insisted they never filmed the front of the house, or its interior. He didn't want people to see the artwork or the antiques, or to be able to easily identify their home when driving by. There had been several very dramatic thefts of valuable items from stately homes and country mansions in the West Country in recent years, and he didn't want to tempt fate by showing off what they had.

In fact, he had recently sold the Roy Lichtenstein canvas for a vast profit, as it didn't really fit in with the rest of the house, and his insurance company had insisted on burglar alarms and outside motion sensitive spotlights being installed to put off would-be thieves when he had had his remaining art collection valued.

During 1989, the BBC had given Will his first solo show, *Forman on Vegetables*. It was a four-part series, each episode an hour long, covering the four seasons in the life of an organic vegetable garden.

This had been shown in the February and March of 1990, before the Chelsea Flower show, and David had produced a book to follow-up on the series. It had been very popular. Not only had it shown people with gardens how to grow vegetables, but he had included window boxes and other containers that people inside apartments and small flats, or with just a balcony, could use to grow their own food. Garden centres around the country had been flooded with people of all ages wanting troughs, pots, buckets for mushrooms and loads of seeds and soil, and this successful series had resulted in the Sunday Times asking to do a major profile of the gardener for their weekend magazine.

Will had discussed this offer with David, and his BBC colleagues, aware it might affect his standing with the gardening community, however unfair that might appear. They couldn't write about him without mentioning David, or rather, Connor Lord.

David agreed to the interview, and the BBC felt it was worth the risk. And, as David said, 'If they turn on you, or turn you off, I've still got a bit of cash in the bank. We won't starve.'

The conservatory had arrived just before the Sunday Times team. The newspaper's journalist, Laurence Hobbs, turned up on the sixth of July, along with a photographer, Peter Vernon; they had been invited to stay for the weekend, to really get to know Will and David, and to take their time when it came to the photographs. It turned out that both men were gay, and they had a brilliant time. When they arrived, their hosts made a big jug of Pimm's and they sat in the conservatory, getting to know each other.

Laurence, who was thirty-one, lived just outside Reigate with a man who bred gun dogs, mainly Labradors and Spaniels. They had been together for five years and were, he said, very content. Peter, who was twenty-nine, lived in a flat near Earl's Court and was happily single.

David asked, 'How did you meet your partner, Laurence?'

'From an ad in *Him* magazine. I was living in Clapham at the time. We chatted over the phone for a month then met up and that was it. We clicked.'

Peter had rolled his eyes, so Laurence asked, 'What's wrong?'

'You taking a month to get laid! Fucking hell.'

'You don't use ads?' asked Will.

'Na, I'm old school. I love the hunt. I love going to a bar or a sauna, seeing a fit young guy and reeling him in.

Ten minutes, tops. If he's not on his knees in ten, I'm off to find a more willing guy. I'll fuck 'em at home, or in the pub toilet, or in a dark room. And, by the way, I always play safe, I'm not that stupid. But I love the chase. And getting my prey! I don't want a relationship. Not yet. Maybe I never will. There are so many cute guys to shag.'

'Do you have a type?' asked Will.

'A lad between eighteen and twenty-five. Scallie lads or trackie boys. Black or white. With tight arses and eager mouths. Show me a lad in a Ben Sherman shirt and pair of bleachers and I'm on him like a shot. I'm a bit of a slut like that. Sorry if it shocks you guys.'

David laughed. 'It doesn't shock us, but it's not what we want. When I started, back at school, you took your pleasure whenever it was available. You couldn't pick or choose.

Same in Oxford. Men who liked men were so rare. But I always got to know them first, even if it was just for sex. There were places for your sort of casual encounter but I never went there. Parks, under bridges, certain back alleys. What we now call cruising areas. There, men, again, took what was available on the night, because they never knew if they would see that person again. But the dangers of being caught, and imprisoned, were too great. Certainly for me. But I always wanted a long-term relationship anyway, even if it seemed impossible at that time. I wanted to find a man to love and live with for the rest of my life, and I did.'

'Because I got drunk,' added Will.

Peter laughed. 'Yeah, that works too. Getting them well drunk or offering a boy a line of coke.'

'Well, I got arrested first.'

'What? Oh, you have to tell us the story,' said Laurence.

'Okay, but it's not for the magazine,' said the gardener. He told them what had happened and they laughed.

'That's really sweet,' said Laurence. 'Guys our age just can't appreciate how difficult it must have been. How frightening it must have been.'

David nodded, 'It was. I didn't offer Will the job because he was gay but because he was a great gardener. But after he started, I fell for him, big time. But if he hadn't gotten drunk, I would never have made a move. If I had got it wrong, if he wasn't like me, my whole life would have been over.'

Will reached over and took his hand. 'But I wanted you too, and still do.'

They just looked at each other, and their guests saw how deep their love still was. Laurence sighed.

'I wish I could put all that down. It's a beautiful story. Will you ever write about it, David?'

'Doubt it. It's private. And to be honest, it might prove a very dull book. No drama, no arguments, no infidelities. Just two men living a happy life together for year after year. My readers would die of boredom. They expect more from me.'

'Like the end of the world or mass riots at football matches?' said Peter.

'Exactly.'

Over dinner that first night, they discussed everything from the AIDS crisis to coming out, mutual friends and the political situation. The two visitors loved the house and were given the full tour; David repeated his demands they didn't take pictures inside, which they accepted reluctantly, but he agreed it would not be a problem to use the conservatory for photographs. And Albert, of course.

Peter had been surprised to find a big white duck wandering around the house when they arrived, and scratched Albert's head when told he was friendly. Albert had followed him about ever since, even sitting on his lap after dinner.

'He's gay, isn't he?' muttered Will, seeing the two bond.

Peter had winked at him and said, 'Told you I like to pull quickly...'

Laurence's article focused mainly on Will, but made it clear he lived with Connor Lord, the author. However, David's name, and their address, were never mentioned. All letters to Will, since that first ever contribution to *Gardeners' World*, had always gone to the BBC first, to protect his privacy.

Peter took one sweet picture of Albert asleep, cuddled up with Henry, in the dog's bed. That went into the magazine, the only interior shot that met with David's approval.

The photographer took a lot of photographs of Will; in the greenhouse, tending the flowers in the herbaceous borders, clipping the *Ericas* in the heather walk, and took one photograph of the two of them together, sitting at different ends of the bench by the pond, with Albert sitting between them. Henry was laying under the bench on his back, tongue out, looking slightly insane.

It was a gorgeous picture and they had a copy framed.

When the magazine appeared later that autumn, that was the picture the editor used for the cover. Under the title, *At home with Will Forman*, the article, six sides long, was funny, warm, full of gardening tips and details and made it clear the TV gardener was living a very happy life as a gay man with a boyfriend, a dog and a duck.

The article didn't affect Will's popularity with his audience and, in fact, the show saw an increase in one group of viewers, older gay men. He might not have been the youngest or hottest man on television, but he was still very handsome, and had a certain charm and warmth that appealed to all age groups.

When they went out in London afterwards, Will got a lot of offers to go home and have sex from his gay fans, which, as usual, he turned down, albeit very politely. He wanted to keep his viewers happy.

Chapter Twenty-Nine

Even though Lottie had retired, confident her "boys" were going to be looked after properly by Nicky, she popped round to see them once or twice a week, and Charlie spent that summer at Wisteria House, working alongside Cal and Will, and playing football with David. They put up a basketball hoop over the garage doors, so he could keep fit, and the four men often played matches there, cheered on by Cal's little girls.

Charlie had just finished his degree in English at Oxford. He had got in aged seventeen and won an Oxford Blue playing rugby his first year. However, a knee injury stopped him pursuing a career as a professional sportsman, so he had turned to writing, songs in particular. That June, in the week after graduating, he had recorded his first album, including a cover of his mother's last big hit, *Take good care of my heart*.

He was waiting for the record to be released in September, and needed to earn some cash, so helped out in the garden, and as a part-time barman at the pub.

The record company hadn't offered him money up front, but good royalties if the album sold well. Morris Holt had negotiated the deal for him. And, as he wasn't twenty-one yet, he couldn't touch his mother's money either, so he had to work.

He stayed with Lottie, but spent a lot of time at Wisteria House, listening to music, playing the piano and "hanging out" with David and Will.

The two men still went to the Fish and Trumpet every now and then. Barney Young had retired and his son Todd was now the landlord.

Barney would sit on the customers' side of the bar, drinking away his pension, happy to have someone else deal with any drunks, and change the barrels of real ale. Todd was happy to employ a fit lad to run up and down the cellar steps doing that for him for a while. One evening, shortly after Will's birthday, they were having a pint there and were talking to a couple of the farm workers. Charlie was behind the bar, chatting up a very pretty girl in a white tennis dress. Will had been negotiating a delivery of autumn manure for September with one of the young farm labourers when David noticed that Neil Strutt, the other lad, was looking very down.

'Neil? What's the matter?'

Neil sighed. He was twenty-two and had just got engaged.

'Went to the housing office to check on where we are as far as a council house is concerned. We've got no chance of buying around here now, cos of the prices, but we'd hoped for a flat at least. Do you know, over two hundred of them council places in Studley Combe have been sold off under the "Right to Buy? Two hundred! And the waiting list for the rest is over ten years. Unless we have a baby. I want kids, but not yet, and not just to get a council flat.'

David frowned, 'Is it that bad?'

'Yeah. It is. All eight of the council places in this here village were bought by their tenants years ago, and we can't buy on the open market. Remember the last cottage that went on the market, on Church Lane? Two up, two down, in terrible nick. Started at one hundred and ten thousand. Went for two hundred thousand quid, to some git from London. We can't compete with that sort of money. They outbid us every time, us locals. And it will be empty most of the bloody time.'

'I am so sorry.'

'Not your fault. I know you own a couple, but you rent them to locals. Not posh wankers from the city.'

As they walked home afterwards, David started composing a letter in his head for the Guardian, a better recipient for this one than the Times, he felt. It ended up as a long opinion piece, meticulously researched, looking at the problems of rural poverty, competition for homes with second home buyers, and the devastation caused by the "Right to Buy" legislation brought in by Thatcher. That law meant that the tenants of council houses could buy them at a discount after living there for three years or more.

The money councils got from the sales had to go into the common local tax pot and could not be used to replace the lost houses and flats. The result was a massive shortage of homes for those in dire housing need. Also, the best houses went first, leaving the poorer stock. Thatcher had been at war with many local councils and wanted to break, or poison, their relationships with their voters, and had wanted to increase the number of homeowners, assuming that they would all vote for her.

David's comments, published as Connor Lord, ripped into her policies, laying out clearly and ruthlessly how she had failed so many of the country's citizens. It finished by saying, *"Her three governments have done nothing for the low paid and unemployed. Poverty is poverty, whether it is found in Birmingham or Puddletown, and the consequences of her policies have condemned many people to homelessness, with housing out of reach for years to come. She should be ashamed of this record, but I doubt she even understands or knows, surrounding herself with "Yes" men, who only tell her what she wants to hear. Or those who benefit from her policies directly.'*

The Labour leader, Michael Foot, raised the article at Prime Minister's Questions, but, as usual, he fluffed the topic. Thatcher stood, ignored the main issue and replied, dripping with contempt, "This government has increased the number of homeowners dramatically, and I will not accept any criticism from the Leader of the Opposition or a writer of cheap paperbacks."

The House howled. The Speaker shouted for calm then let a Labour backbencher speak, instead of the Leader of the Opposition. He called out; "Brad Smith". Smith stood up. He had been briefed by the editor of the Guardian.

'Mr. Speaker. It is well known that the Prime Minister has little time for recreation, so it is hardly surprising that she has not had time to read any of Connor Lord's books. Far from being cheap paperbacks, they are well researched novels that have raised important issues and questions over the years that we should all be considering, just like this recent article about housing and rural poverty. That the Prime Minister should be so contemptable of a man who has brought millions into Britain through taxes and book sales, and who has relentlessly focused on the damage we are doing to our environment, and who exposed the nonsense of her policies regarding gay rights, is shameful.

The truth is, he is right; her policies have hurt the poor and the weakest in British society and continue to do so. She should resign.' There was uproar in the House, and Thatcher stomped out, fuming. She made it known through various channels that Connor Lord was persona non grata as far as she was concerned, which didn't bother David one bit. He had made his point.

The fact that it didn't do any good, or lead to any changes, was disappointing but at least the issue had been raised and the Prime Minister's nose put firmly out of joint.

On the 28th of November, Thatcher was removed from office by her own colleagues, terrified by the growing hostility towards the Conservative Party due to her arrogance. They had already had to contend with the Poll Tax riots back in March, and the resignation of her Foreign Secretary, Geoffrey Howe, over her attitude to the EU. Enough was enough.

David and Will were glued to the TV as she left Number 10, Downing Street in tears. They were drinking champagne. To add insult to her injury, John Major, her surprising successor as Prime Minister, recommended a knighthood for David in the New Year's Honours list, for services to charity and publishing. No one dared ask Mrs Thatcher what her views were on this honour, fearing the response might give her a heart attack.

*

David's investiture took place at Buckingham Palace at the beginning of March 1991. He and Will took Lottie along with them, and the three of them stayed at the Savoy the night before, to give her a treat. They waited in the queue for him to be invited to kneel before his sovereign, who tapped him on both shoulders with her sword, and told him how much the Duke of Edinburgh enjoyed his books. He managed a quick, "Thank you, Ma'am" before an usher swept him back to his guests and the slow procession of newly honoured people leaving the Palace. Lottie loved every moment, and took great pleasure in telling Ginny about it, several times, when they got home again.

Just before Easter, on Maundy Thursday 28th March, Lottie went to buy some hot-cross buns from the bakery at eight-thirty in the morning. As she stepped out of the shop, she stopped, gasped and fell to the pavement, stone dead.

She had had a massive heart attack. She missed seeing Charlie's first performance of his number one single on *Top of the Pops* that evening. The producers had to show a video of him singing, as he was too upset to perform live.

David and Will were devastated by her death. She had been with them right from the start and had become a substitute mother to both, as well as a very close friend.

Lottie was buried the following Friday next to her daughter in St. Stephen's churchyard. David gave the eulogy, and Charlie sang *Amazing Grace* during the funeral service, and stayed at Wisteria House for a week afterwards, as he couldn't face being in the cottage on his own.

During that week, he and David had a serious talk about his financial situation. Thanks to Morris Holt's contract, he was making good money from both his single, and the best-selling album it was taken from. He also now owned Lottie's cottage and, in four months, he would get his hands on Flo's money as well.

Sitting in the study, Charlie said, 'You're rich, David, but you don't flash it around. I like that. Always have. You're cool with money. I want to be like that too. What should I do?'

'Well, first, decide where you want to live. I assume it is not here.'

'Yeah, I want to be in London.'

He was living in a shared house in Brixton at the time, with two men and a girl who worked for his recording company.

'Okay then. Buy a house in London. Prices are rising again, and they will keep on going up, so it would be a good investment, apart from being your home.'

'I want my own recording studio too.'

'Then buy a house with a basement and build one there.'

'Cool. I like that idea.'

'As far as Lottie's cottage is concerned, well, we did a bit of work on it last year, so it is in good nick. If you sell it, some knob from London will buy it as a weekend place, a second home. I'd keep it and rent it out to a local instead. You might want to have a place in the country yourself one day, when you're married with kids. Keep it, rent it to someone who lives locally, help a young family stay in this area. In fact, I know of a young couple who are in desperate need of a place and they are very sensible and would be good tenants.'

He was referring to Neil Strutt and his fiancée.

'I like that idea too. Yes. I'll do that.'

David grinned and rubbed his hands together. 'Okay. We're on a roll here. You need to save and invest money, just in case people stop buying your music, or your next album is shit. I'm sure it won't be but you never know. Buy shares in high-tech companies like Apple, or maybe Microsoft, as they are going up and up, and some long-term government bonds. I've got a guy in the City who gives me good advice. He charges, of course, but he's not greedy. But keep your money under your own management.

Pay for specific advice, but don't let some slick City company get their hands on everything. They charge shockingly high management fees for doing fuck all. You can manage it yourself. You can always ask me for advice and that would be free.'

Charlie had not looked too convinced, so David took him to his computer and showed him his investment spreadsheets. Charlie gasped and said, 'Christ, you're worth a fucking fortune.'

'Don't tell anyone, please. Which is my final bit of advice. You will discover, especially if you're in London and in the music business, that you will be surrounded by people who want to be your friend. Hangers-on. They will ask for money. Don't give it to them or tell anyone how much you're worth. If you want to help poor people, give to a charity. We give a lot to AIDS charities and "Save the Children", as their work is vitally important and their overheads very low. Some charities spend far too much on their management and don't give enough to the people they are meant to be helping. Give some cash to the homeless, or a lad or a lass begging on the streets, but not your mates.

They will never give it back and you will get the reputation for being a soft touch. Oh, and get yourself a good accountant and always pay your taxes. People, even your fans, hate tax cheats. And, needless to say, don't do drugs. I know you don't, but the temptation will get worse as you become more and more successful.'

Charlie nodded. 'I know, I've been offered loads already. I've never touched any, not even a cigarette. Don't drink either, much. I never knew my mum, just from her records and stories from Gran and you two. I missed out on having a mother due to that shit. I'm not interested in starting.'

David had given him a hug and said, 'So glad to hear that. You'll do fine, Charlie, and we're always here if you need advice, or just want to chat, or if you get into a mess.'

'I know. You are my two dads, you two are.'

He ended up buying a house in Warwick Gardens with a vast basement, which he turned into a recording studio. He started dating a girl from the estate agency that sold it to him, a very sensible, down-to-earth girl called Cassie Hart. She hated the drug culture, made sure there were very few "hangers-on" around him, cooked amazingly healthy food for them both, made him go jogging, and was the inspiration behind his third hit single, *She's my girl*. David and Will adored her from the start.

She and Charlie got married the following year.

2000

Chapter Thirty

Fletcher's Cross, 16th July

'Have you seen my glasses? I'm sure I left them on the table in here.'
'They must be by the stereo. You were wearing them when you put Charlie's cd on.'
'Oh, right. Yes. Thank you, love.'
David went back inside, found his reading glasses then returned to the conservatory where they were having a very long and peaceful breakfast. Charlie's latest album was playing softly in the background, a collection of love ballads, accompanied by a single guitar. It was number one in the album charts, and he was now a best-selling artist, right around the world. He still dropped in from time to time, with his wife and two children.
As David sat down again, Will asked, 'You're not going a bit do-lally, are you, love?'
'Not yet, William. We've both done rather well when it comes to health. But I bloody hope my brain isn't going. That would be dreadful. Now, where's that review?'
David picked up the Sunday Times, opened it at the television review section and started to read aloud.
"*It is over thirty years since the BBC's adaption of the John Wright stories, and this new series, written by the author, Connor Lord, himself, is a triumph; of production values, of acting and of script writing. ITV has pulled out all the stops, recreating a post-war Britain that reeks of the smoke, fire and dust rising*

from the rubble of a bombed-out London. The story, of spivs and war racketeers, raises comparison with the rapacious nature of bankers and modern investors, earning money out of desperation, job losses and factory closures, making it so relevant to a new generation of viewers."

David read on then said, 'He loves the acting, the clothes, the cars, the lot. Good.'

Will grinned. 'Another success.'

'Indeed. Hope it doesn't upset the BBC too much.'

He still had friends at the BBC. They had turned *This Poisoned Land* into a six-part series in 1998, which had won several BAFTA awards for acting, special effects and design. The scenes set at the football match when the crowd had gone insane had become TV classics.

The gardener grunted at this comment. 'Well, fuck 'em if it does.' Will was still smarting after just being let go by the corporation, who had wanted to update their venerable gardening programme. He had been replaced by two young bouncy presenters on *Gardeners' World*, an Anglo-Indian girl and a young black guy from Brixton, in an attempt to bring in younger, more diverse, viewers. It had totally failed, and the BBC had taken on Alan Titchmarsh instead. Will still had his column in *Country Life*, but his glory days on British television were over. He was sixty-five, and had been forced into unwanted retirement, and was still sulking about it.

'Yes. Fuck 'em for the way they treated you. Those new chaps are rubbish.'

'Chaps?'

'I'm still reliving the 1950s after last night's drama.'

The revival of the John Wright books had come about due to another retirement, Murrey Holt's. When the rumour mill started, that Holt was giving up, several major agencies tried to pitch their services to the author. But he had gone up to London and sat down with Murrey's son and told him that, if he wanted, he would be happy to stick with The Holt Agency. Morris Holt, greatly relieved at not losing their biggest client, had suggested it was time to revive David's old spy books, long out of print.

Morris contacted Gloria Hilton. Hilton Books had merged with an American company, Lincoln House, back in the nineties, then bought up several smaller publishers, in England, India and the States. Gloria was now the CEO of H&H, the Hilton House Group, the biggest global publishers, and was based in New York.

However, she was well aware of the debt she owed to Connor Lord. She listened to Morris's idea; the republication in paperback of all ten Wright books, with stunningly designed new covers, in a retro style. She went for it, and they came out in 1997. They were a huge success, as a new generation discovered one of the original spies.

ITV had bid for the new TV rights, then suggested David produce the screenplays himself, with the help of an assistant experienced in script writing. She duly arrived at Wisteria House, stayed for a month, and Penny Kelly was credited under Connor Lord's name for the adaption of the first book. While he had appreciated her skills, they hadn't bonded very well. She was very left-wing and sneered at their lifestyle. Thereafter, David had worked alone. Now that first story had been shown. He reached for his laptop, opened it at his "Connor Lord" website, and started reading through the emails that were piling in. He read, nodding and smiling. They all loved the new series.

Will was happy for his lover, despite his own recent disappointment. He sat back, relaxing, sipping his coffee and looked up at the glass roof above them. Big thick succulent ferns, called stag horns, hung from terracotta pots suspended from the rafters, and the air was filled with the scent of lemon blossom. The doors to the terrace were open and Henry the Third, a black Labrador, was laying in the sun, sleeping peacefully.

They had gone back to a Labrador, mainly because Laurence Hobbs' boyfriend had a bitch which produced a litter, and they liked his dog.

They had kept in touch after his interview, visiting the kennels on one occasion after returning from a trip through Gatwick Airport. David had wanted to call the new dog Flush Two, but Will put his foot down, ignoring their "no conflict" rule, saying, "I'm not standing outside in the middle of the night shouting Flushed, Flushed…"

The second Henry had died young, of cancer, in 1998. Albert had followed him to his grave shortly afterwards, at a very great age for a duck. They weren't sure why but they suspected he died of a broken heart.

It was a beautiful sunny Sunday morning at the height of summer. Will breathed in deeply, enjoying the fragrance of his trees.

'So, we're not going to church then?'

'Do you really want to, after last Sunday?'

'No.'

James Fallow had finally retired in 1993, and they had had three new vicars in the last seven years. The first had been a drunk, the second a bore and the new one, who had started the week before, a slap happy born-again Christian, who had asked everyone to shake hands with their fellow worshipers and shout "Hallelujah" after each prayer. It had been, in their eyes, a total fiasco.

'Praise the Lord.'

'Hallelujah.'

They both laughed.

David looked at Will and said, 'Have I told you I love today?'

'Not yet.'

'Well, I do. Very much indeed. As much now as I did when we first got together. More even if that is possible.'

Will got up, came around the table and gave him a hug.

'Right back at you, old boy.'

'A little less of the old, thank you very much.'

'You're sixty-nine, love. We're both getting ancient. Do you want some more coffee?'

'Yes, please.'

Will picked up the cafetiere and disappeared inside. The doorbell rang. David got up and shouted, 'I'll get it,' then mumbled 'who the hell is that at ten on a Sunday morning' to himself. It was a delivery man from the Interflora shop in Studley Combe. He was holding a huge bouquet of flowers.

'Sir David Manners?'

'Yes, that's me.'

'These are for you, sir. Have a nice day.'

David carried the bouquet into the kitchen and laid it down carefully on the table. Will gasped and said, 'Bloody hell. Who sent you those?'

'Not sure. There's a card.'

'Can you read it without your glasses?'

'Fuck off,' was the reply, but he was laughing as he held the card as far from his eyes as possible and squinted at it.

'Oh, it's from Simon at Royce Productions. Congratulating me on the programme. That's nice of him.'

Royce worked exclusively for ITV and had been the company that filmed the Wright story.

'It is,' said Will, reaching into a cupboard for their largest vase. He filled this with water, then looked at the flowers.

'Nice bouquet. White and yellow roses, white Sweet Williams, pink and white carnations. It's a bit like a huge wedding arrangement.'

'Can you sort them out? You're much better at doing that than I am. I'll make the coffee.'

He ground the beans and added boiling water, as Will unpacked the flowers and arranged them in the engraved Lalique glass vase, then carried it through and put it on a mat on the piano. His lover, bringing the coffee back, said, 'Oh, nice job. They do look splendid.'

The music had stopped, so Will asked, 'What do you want to hear now?'

'Enya?'

'*Watermark*?'

'Oh yes please.'

They both loved the singer's second album. David had taught himself how to play *Orinoco Flow* on the piano, and Will would often sit next to him on the piano stool, watching his fingers produce the much beloved song, before giving him a kiss and saying, 'That was great.'

Having poured the coffee, David sent an sms to Simon to thank him for the flowers and settled back to finish the Sunday papers. After a while, Will said, 'It feels a bit odd, not going to church. What shall we do instead?'

'Well, we still haven't decided what to do for lunch. What do you fancy?'

'How about a walk around the grounds of Langham House, then lunch there. They have that summer buffet thing going on outside on the terrace at the moment.'

'Excellent idea. They don't mind us taking Henry, do they?'

'Nope. Dogs, well-behaved ones that is, are always welcome. I'll call and book a table.'

Two hours later, as they were walking through the topiary garden back to the hotel, Will stumbled. He didn't fall, but swayed a bit and David held him. 'Are you alright?'

'Yes. Yes. I just felt a little dizzy, that's all.'

'How strange. Have you felt that before?'

'No. No, I'm fine now. Don't worry.'

David looked very concerned. Will was never ill. Apart from the occasional cold, they had both survived thus far without any major health problems.

They reached the terrace and were led to their table under a blue umbrella. David ordered a bottle of Chablis, before they got up again and fetched the first round of food from the long buffet. Smoked salmon, devilled eggs, and small pots of an excellent prawn cocktail.

The terrace was filling up. With couples, small family groups and one big party of ten, celebrating a grandmother's birthday. They chatted about nothing in particular, then went back to collect their second course of cold roast beef, potato salad, beetroot salad, warm bread rolls and some cheese. The hotel's catering was very popular, and very good, and they finished with two crème caramels and coffee. Henry lay peacefully under the table throughout the meal. He was a very well-trained animal.

That night, at about eleven-thirty, David was woken by a feeling of dampness in the bed. He felt around him and realised the bottom sheet was soaking wet. He sat up, turned on the bedside light, and saw Will staring at the ceiling, struggling to breath. He had had a stroke and wet the bed. Trying to control his panic, David called for an ambulance, then called Cal, as he was totally unable to drive to the hospital himself, so shaken was he by this event.

The ambulance came in seven minutes and drove Will away to Stroud General. David stripped the bed, put the sheets in the bath, took a quick shower and dressed. He came out of the back door to find an equally shaken Cal waiting by the car.

'Is he going to be okay, David?'

'I don't know. I don't know. He's still alive at least, but he couldn't speak.

I'm not sure if he could even hear me.'

'Oh shit. Come on, let's get there.'

He drove fast but not too fast and they reached the Casualty Department of Stroud General Hospital in twenty minutes. There was a very officious nurse there who demanded to know what David's relationship was with Mr. Forman.

'I'm his family. We've been together for over forty years.'

'But you're not real family, are you, dear. I can't give out information to anyone other than real family.'

Cal and David were both shocked. Luckily there was a senior doctor on duty, who recognised David, being a big fan of his books. He came up and said, 'Nurse Jones. Try not to be a total arsehole. This is Sir David Manners. Better known as the writer, Connor Lord. He and Will Forman are about the most famous gay couple in England, after Elton John and his boyfriend. Of course, we can give him information. Please, Sir David, follow me.'

He led them to a side room. Will had been checked in the emergency room, then transferred here. He was in bed, surrounded by machines, unconscious. The doctor said, 'It looks worse than it is. For now. He's had a mild stroke, that's affected his right side, which means the blood clot was on the left side of the brain. We have carried out thrombolysis with an injection of alteplase which had cleared the blood clot. It was a mild ischemic stroke, which are the most common. He is breathing on his own but has some mild loss of muscle function on his right side, as I said, which means he might need rehabilitation, depending on how severe the damage is. We'll know more in the morning. But so far, it looks hopeful. Has he had any symptoms before?'

David, still in a daze, asked, 'Such as?'

'Dizziness, loss of facial movement, that sort of thing.'

'He felt dizzy just before lunch but otherwise he's been as healthy as a horse.'

'Was he sick? Vomiting, I mean. Earlier this evening?'

'Not that he said, but he's a proud chap. He might not have admitted it.'

'I see. Well, he was slightly sick in the ambulance, but it was just fluid, no solids.'

David sat down heavily in the chair by the bed and burst into tears. He had never cried in his life before, but now, threatened with the loss of his lover, he was overwhelmed. Cal put his hand on his shoulder and said, 'It's okay, David. It's okay. He's alive and in very good hands.'

David looked up at the doctor and said, 'Can I stay?'

'Of course, you can.' Turning to the gardener, he asked, 'Are you a relative?'

Cal shook his head. 'Close friend.'

'Well, if Sir David is staying, why not go home and get a good night's sleep. No point in both of you sitting here. Mr. Forman won't wake up again tonight. He been heavily sedated.'

Cal said, 'I don't mind…' but David squeezed his hand and said, 'Get yourself off home, Cal. Can you let Henry out and give him his breakfast in the morning? And tell Nicky what's happened. I'll sort out the bedroom when I get home later.'

Cal nodded, gave his old friend's shoulder one more squeeze and left. The doctor went off to see another patient, and David sat, holding Will's hand. Tears ran down his face, but he didn't wipe them away. He felt so useless.

Will just lay there, sleeping, as the machines monitored his heartbeat,

oxygen levels and respiration.

Eventually David fell asleep in the chair and the night nurse slipped a blanket around his shoulders, checked Will's vitals and left the two of them alone.

Chapter Thirty-One

Will was in hospital for five days. By the time he came back to the house on the green, David had replaced the bed and found a nurse who could help him regain his full movement though rehabilitation exercises at home.

Cal had got a friend of his with a van to help him take the old bed to the rubbish dump, and a call to Harrod's Bedding department had resulted in a new Swedish Hästen's bed being delivered the next day, along with a very thin rubber under sheet in case of future accidents, new pillows, a big continental quilt plus cover, and new cotton sheets. For once, David used his title, and his Connor Lord name, and lots of money, to solve a problem very quickly.

Will had woken at seven in the morning that first day in hospital and simply said, 'Where am I?'

His voice had been slightly slurred, and one side of his face didn't move properly. David had hugged and kissed him, then explained what had happened, leaving out the bed-pissing bit.

'Will I be okay or an invalid for the rest of my life?'

'The doctor is very hopefully you'll get back to normal.'

'When?'

'Soon. Well, I don't know to be honest.'

'Will I be able to move again? I mean, I love growing vegetables but I don't want to be one.'

David laughed. The old Will was still there.

'You won't. We'll get you back on your feet. And on your back again.'

His lover studied his face and said, 'You look exhausted.'

'You look lovely.'

'I'm not sure I could give you a blow job at the moment. My mouth feels all funny.'

'Well, the nurse will be in soon so we better hold off on the sex for now.'

'If you're sure, Davey.'

'I am. Just relax and get better. By the way, were you sick last night?'

Will thought about this then nodded, very carefully. 'I was. I felt embarrassed to tell you as we'd had such a lovely lunch. Why?'

'It might have been a warning symptom, like the dizziness.'

'Oh. I didn't know that.'

'Neither did I.'

The nurse had arrived, then a new doctor, who confirmed Will was on the mend. David had taken a taxi home and started the process of sorting out the bedroom, ready for his partner's return. He also called Charlie Nolan to tell him what had happened. When he went back to the hospital at five o'clock, there was a huge black man in a suit standing outside the private room. This was Percy, Charlie's bodyguard. Cassie Nolan refused to let her husband have a posse, but such was his fame at that point, he did need protection. There was a buzz of excitement David had noticed when he arrived at Stroud General, and now he knew why.

'Afternoon, Percy.'

'Afternoon, Sir David.'

Percy was very strict about using his title.

'Can I go in?'

The huge man shrugged and stood aside. Cassie stood up and hugged him as he said, 'Is it possible Percy is even bigger than the last time I saw him?'

'Yeah, he's a growing lad. Hi, David. How are you doing?'

'Better than this lazy sod. Look at him, laying in bed when the lawn needs mowing.' He bent down and kissed Will, who said, 'I can move my face again. Look.'

He grinned and both sides of his mouth worked, sort of. David smiled and said, 'That's brilliant. Hi Charlie.' and stuck his hand out and squeezed their boy's hand across the bed. He added, 'Nice of you to come.'

'Of course I'd come. You're my dads.'

They sat and talked for an hour but then the singer and his wife had to leave as he was flying to the States that night.

'Give our love to the kids. And thank you so much for coming.'
There were hugs all around then they left. Outside the room there
was a huddle of nurses waiting to see them as they walked out, back
to the black Range Rover parked outside.

David went in twice a day until his sick lover was allowed to go
home. Cal helped him get Will upstairs when he came back. He
could walk, just, but dragged his right leg, as he couldn't lift it
properly. His right arm was weak too, and if he tried to raise it, it
fell down again. They got through the first night together without
any problems, but, in the morning, Will called out from the
bathroom.

'Davey? I'm in trouble.'

David went in. His lover was naked and sitting on the lavatory but
leaning against the washbasin.

'I can't wipe my arse. Every time I try, I fall over.'

'Well, I'll just have to do it for you.'

Will went red and shook his head.

'This is so humiliating.'

David knelt down in front of him and hugged him.

'William Forman. When I think of all the pleasure and joy I have
had from your bum over the last forty-three years, it is the very
least I can do.'

He kissed him and said, 'Lean against me,' as he reached for the loo
paper. Job done, he helped Will sit on the edge of the bath, then
showered him, dried him and led him back to the bedroom. He
helped him dress in a white T-shirt and jeans and brushed his hair.

'Still got a nice touch of the James Deans going on.'

'He's been dead for fifty years…'

'Exactly…'

They got downstairs and David sat Will in the conservatory and
went to make breakfast. At ten the nurse arrived for Will's first
rehabilitation session.

His name was Robert Harrison, or "call me Robbie". He was being
paid for through their private medical insurance. He told them he
was thirty, and ex-Royal Navy, where he had served as a medical
orderly on the aircraft carrier Ark Royal; he had retrained as a
rehabilitation therapist after leaving the service. He sat with Will in
the conservatory and said, 'Right then, let's see what you can do.'
He went through Will's capacity to move, noting the level of
disability, then said, 'How do you spend most of your time

normally?'

'In the garden or in the greenhouse.'

'Right. So, let's walk there and take a look.'

Will had frowned and said, 'Aren't you going to make me do exercises then?'

'Yes, but in a way you will find comfortable. Come on.'

The hospital had given Will a walking stick and the two of them had made their way slowly across the lawn and into the greenhouse. Robbie then worked out a series of movements for Will to practice, including moving flowerpots around, kicking against bags of potting compost with his weak leg, and sitting and standing up (slowly) and reaching up to touch the grapes hanging from the roof. He couldn't grip things properly with his right hand, so just slid the pots about on the bench to begin with.

Back in the conservatory over coffee, Robbie said, 'Right then. I'll be here three days a week to keep you motivated. The rest of the time you should do those movements on your own. With the meds, and these exercises, you should be back to normal in about three months. Maybe less. You have been lucky. It wasn't a bad stroke.'

'Seems bad enough to me.'

'You're still alive, aren't you?'

Will had laughed, 'You're right there.'

David had joined them and now asked, 'Could he have another?'

Robbie nodded. 'It's possible, but the meds are good and getting better. Once he's fully recovered, a simple aspirin a day might be good enough to keep you from having another one, Will.'

'Good. So, what's your story? Where are you based?'

Robbie smiled. 'I live in Stroud with my boyfriend. He's a hairdresser.'

'Ah, you're family then.'

'Oh yes.'

'How long have you two been together?'

'Two years.'

'What sort of relationship?'

The nurse frowned and replied, 'What do you mean, David?'

'Open or closed?'

Robbie laughed. 'Oh, I see. Um…'

'It's alright. We don't want to seduce you. We are very faithful and monogamous. Always have been.'

The nurse grinned. 'Well, we play with other guys sometimes. Threesomes, or we pop off for a bit on the side solo.'

'But you're happy?'

'Very.'

David nodded. 'Good. That's nice. I like the way you've designed his treatment to fit in with the things he likes doing.'

The nurse nodded. 'It's the best way. If I tried to give him loads of other exercises they might not work so well as they would be out of Will's normal routine. I find fitting the movements to one's daily life much more effective.'

He left and they had lunch. Will was exhausted and lay down on the sofa and fell asleep. David, a bit overwhelmed by events, went and sat in the garden, trying not to cry again. He had nearly lost Will, and it had shaken him to the very core.

Robbie had been right. It took three months before Will was back to normal, but his therapy worked and he regained full use of his limbs again.

They invited the nurse and his boyfriend for Sunday lunch at the end of the sessions, and really bonded over the roast chicken. Robbie's friend turned out to be a handsome young Indian man called Raj, who was very charming and funny. After that, the two couples became close friends and the younger pair regular visitors to Wisteria House.

Chapter Thirty-Two

During Will's recovery period, they spent more time together than in recent years. Although they had always started their days together and ended them the same way, the bit in the middle had seen them working on their own, David in his study, Will out in the garden. Now they spent all day in each other's company and discovered they enjoyed it immensely.

As Will couldn't drive, David would get him into the passenger seat of their Range Rover and take him off for day trips to nice cafés, or to have a picnic in a secluded spot by the Severn estuary.

Those were becoming rarer and rarer to find. The Cotswolds had become even more popular as a tourist destination, with coachloads of new Chinese visitors adding to the existing numbers of Japanese and elderly Americans, taking endless photographs of thatched cottages and ancient stone churches, flower-filled cottage gardens and riders on horseback clopping through the villages.

The small farm shop where they had bought their eggs and ducks had grown in recent years to become a huge organic supermarket, with a café and restaurant attached. The owner, the son of the woman who had started the first store, even set up a petting zoo with cute baby farm animals for children to touch and play with, including little goats, rabbits, guinea pigs and chicks. It had become a very popular place for families to visit on Saturdays and Sundays, which was proving a nightmare for their closest neighbours, as the queues of traffic were awful.

There was one nice place close to Slimbridge they loved to go to, which was unknown to the general public, with a picnic table and a bench overlooking the water. They often went there, watching Henry sniff through the long grass and put up the occasional duck, as they drank coffee and ate ham and cheese sandwiches. They would watch the oystercatchers dibbing in the estuary's mudflats, or flying in pairs, whistling to each other as they moved fast along the river. They would sit in silence sometimes, or talk all the time, depending on their mood. The silences were never awkward, and the conversations wide-ranging, from the television programmes they had watched the night before, to films, politics or the new celebrity culture.

Although David had a great imagination, he had never considered the possibility of people becoming famous for simply being famous. The original celebrities they had grown up with had been film stars like Elizabeth Taylor or Richard Burton, or great footballers like George Best. People who had become famous for being excellent at something.

Charlie was famous too, but because he was a best-selling recording artist and a fantastic performer. But now they had "celebrities" like the Kardashians, famous for simply existing and being filmed doing so. They had watched five minutes of one of their shows and had had to turn it off, due to the infantile nature of their conversations. They used a Sky Box these days, which gave them access to a wide variety of channels. It had replaced their original, very discreet, satellite dish, put up in the eighties. That had given David access to CNN in Fletcher's Cross. He had seen it before in hotel rooms, but now he could follow the new twenty-four-hour news cycle at home if he wished.

It was through that dish that he had stumbled across Fox News in 1996.

He had been sitting aimlessly clicking through the channels, when he had come across *The O'Reilly Report*, which would later be known as *The O'Reilly Factor*. David had sat staring as the arrogant, loud-mouthed presenter had shouted at his viewers, giving his opinions as downright facts on various topics, all from a distinctly right-wing bias. He watched it over several days before he called Morris Holt in London.

'Morris? It's David.'

'Hello. What can I do for my favourite author this fine September morning?'

'Have you ever seen a show called *The O'Reilly Report*? On Fox News?'

'God yes, but only once. Thought it total crap. Why?'

'They have pinched my idea from *Here is the News*!'

Morris laughed, thought for a moment, then said, 'Oh yes. It's just like your book. Do you want to sue them for copyright infringement?'

'Could I?'

'No, sorry. It's too much of a general concept for you to win a case. I can run it by the legal department but that's what I think they will say. And if you tried it would only give them much needed publicity. Only a handful of people watch that shit at the moment. A court case would be free advertising for them.'

'It is shit, isn't it? But it's dangerous shit.'

'Seriously? You think so?'

'If it becomes popular, it might be.'

'Well, not sure what you can do about it.'

'Fox is owned by Rupert Murdoch, isn't it?'

'Yes.'

'Well then, I'll do the only thing I can do. I'll cancel my subscription to The Times.' Murdoch owned the Times, the Sun and the News of the World newspapers at that point in time. Morris laughed again. 'I'll do the same then. In solidarity. Apart from the Sunday Times. I need to read the Literary Supplement, of course. But Murdoch's an awful man.'

From then on, they had only had the Guardian and the Mirror delivered to the house on the green. Apart from Sundays. David wanted to keep an eye on business too.

When the weather was bad during Will's recovery period, they stayed inside and watched things from Will's DVD collection. He had all the *Alien* and *Star Wars* films, plus TV series like *Babylon 5*, *Star Trek the next Generation, Buffy the vampire slayer, The West Wing* and *The Wire*. He also had many blockbuster films, like the *Godfather* trilogy and *Titanic*. They had seen that when it came out in 1998 and loved it, apart from the bit when Jack had slipped away from Rose on the raft. Will had turned to David in the cinema and muttered, 'What a bitch. There was plenty of room for the two of them on that door and she let him drown, the miserable cow!'

The first day Will could walk without using a stick, they made their way, very slowly and carefully, up to the beech wood.

They had brought a thermos and sandwiches as usual and sat on the fallen trunk together. They drank some coffee, looking at the tall trees with their grey-brown trunks, the thick canopy of leaves overhead, listening to the birds. A pheasant called in the distance and the cows in Robson's field behind the wood stood chewing their cuds. It was a scene of total peace and classic English countryside beauty. After a while, Will said, 'Do you ever miss smoking?'

'Sometimes.'

'I do. Sitting here, like this, in the fresh air. It makes me want to reach for a packet of Benson and Hedges and light up.'

'It's funny. I was just thinking the same thing. But we'd better not start again. Might be bad for us.'

'It might.'

'But habits die hard, don't they? Even the bad ones.'

Will had nudged him and replied, 'Am I a habit then?'

'You're not a habit, you're my reason for living. You keep my heart beating.'

He straddled the trunk and pulled Will close and sang, 'And, sometimes, all I need is the air that I breathe and to love you.'

Will turned and kissed his lover and said, 'Oh Davey, you old romantic. This is so nice. Being here with you. Walking here again, unaided. I feel as if I'm coming back from the dead.'

'I thought you hated zombie movies.'

'You know what I mean. God, you never forget anything, do you?'

'Very true. I would hate to get dementia and forget my past. Forget you.'

'But would you know you'd forgotten me?'

'Maybe. I don't know. And I don't want to find out either. Nor have you suffer from it.'

'Ta. Are you hungry?' Will had got his grip back and now unwrapped the silver foil and offered his lover a ham and cheese sandwich, in home-made brown bread, carved from the loaf by the recovering man that morning. They ate in silence, watched by a very attentive Henry. His patience was finally rewarded as they both tossed him their crusts. They had made their way home again, slowly as before, stopping to look at the gingko trees and the fish in the pond, the bees and the hummingbird hawkmoths hovering around the lavender flowers, before David sat Will in front of the television and went to cook their dinner.

They repeated that route many times during August and the first part of September, with Will getting stronger and faster with every walk. They also took walks around the village, greeting old friends who asked after Will's health, and one afternoon they took flowers from the garden and put them on Lottie's and Flora's graves. They missed Lottie very much.

Robbie came for his final therapy visit and said Will seemed back to one hundred per cent normal.

They had to wait two more weeks before Will's doctor told them he was fine and dandy and they could start having sex again if they were so inclined. They were…

Chapter Thirty-Three

The evening after the doctor had given Will the all-clear, they were in bed about to have sex for the first time in over three months. It was the longest time in their entire relationship they had not made love. They had lain, side by side, just touching and stroking each other for a while, before David had asked, 'Are you sure you want to risk it?'

'If we leave it any longer, it will be like when you took my virginity.'

'You remember that?'

'Like it was yesterday. I'd waited so long to have sex with a man then there you were, kneeling in front of me as I lay on the bed, legs up, frighten to death but as excited as hell, waiting for you to push in.'

'It was wonderful. You looked so scared but were as hard as a rock. Like now.'

Will chuckled, 'Yes. I've missed having you inside me.'

'I don't want to kill you.'

'You're not that big...'

'Very funny. You know what...'

'I know. If it happens then at least I'll die happy.'

'Okay then. How do you want it?'

'I want to see your face.'

'Are you sure? There are a few more wrinkles than there were all those

years ago.'

'Looks the same to me, Davey.'

His lover kissed him and replied, 'Flattery will get you laid, love. Ok, on your back then.'

'Yes, sir.'

It all worked nicely and no one died. After they had cleaned themselves up, they lay in their favourite position, with Will resting his head on David's shoulder, one arm across his chest.

'Mmm, that was great.'

'It was. I've missed your bum.'

'And I've missed your cock. Still the same. Just as I remembered. By the way, I like this new mattress. It's very comfortable.'

David had finally told him what had happened to the old bed. 'It should be.'

'Oh. Expensive?'

'It needed a mortgage.'

'Well, it was worth it.'

There was silence, then Will sucked in air and hesitated. David knew the procedure when his lover wanted to say something he found difficult and asked, 'What is it?'

'I… I wanted to say thanks.'

'For the sex?'

'No. For you being so good when I came out of hospital. You know, in the bathroom and stuff.'

'You're most welcome.'

'You made it so easy and comfortable for me.'

'You would have done the same.'

'I hope so. I've always found that sort of thing tricky.'

'What, like with changing Charlie's nappies when he was a baby.'

'Yep. Once you'd got over the initial shock of the whole thing after Lottie showed us what to do in the kitchen, you never seemed to mind doing that, however evil the stuff he'd produced.'

'God, some of it was very nasty, wasn't it? I thought we'd have to call in *Doctor Who* some days.'

'The Green Death?'

'Exactly.'

That was the name of a *Doctor Who* story from the seventies. It was also the name they had given to a particularly nasty poo Charlie had had after eating too much spinach. They had often had to change the toddler when Lottie was busy or out shopping, once they had gotten over their horror at the thought.

David sighed and said, 'He needed changing and couldn't do it himself. Same with you. No point in making a fuss about it. It wasn't your fault.'

'Still. I appreciate it. Very much.'

David kissed his head and hugged him closer.

'I thought I'd lost you. It was terrible.'

'Well, I'm still here, and I'm going to do everything I can to stay around.'

'Please do. I don't think I could go on without you.'

'Me neither if you died.' He then frowned and asked, 'Hear, where would you have buried me?'

'At St. Stephen's. I've bought the plot next to Lottie and Flo.'

'Really? You never told me that.'

'Well, it's good to be prepared. The graveyard's filling up. I wanted to make sure we got in.'

'But you said one plot? Where are you going to be buried?'

'On top of you.'

'In death as in life then?'

'Yup.'

They both laughed and laughed. Will, gasping for air, said, 'Oh, you're a silly sod at times, Mr. Manners.'

'Hmmm.'

'Had you picked out the music too?'

'For your funeral? No, I hadn't. Got any preferences?'

His lover considered this question carefully. 'Well, I love Pachabel's Canon. You could play that while people are waiting for the service to start.'

'That's usually a wedding tune.'

'Well, we'd have missed having one so that's no problem. Oh, or you could play that song from *"Robin Hood, Prince of Thieves"* I liked.'

'The Bryan Adams one? *"Everything I do I do it for you"*?'

'Yeah, that one.'

'Christ, I'd be crying right from the start.'

'You would? You never cry.'

David was silent. Will looked at him and touch his chin. 'Did you cry when I was in hospital?'

'Mmm.'

'Oh, Davey, I'm so sorry.' He kissed him on the chin.

'I thought you'd gone.'

'I'm still here. Oh, I know what you could play when they are carrying out the coffin.'

'What?'

'*Spirit in the sky*. The original version by Norman Greenbaum. Not that shitty one by Gareth Gates.'

'Okay… bloody hell. They'll all be dancing out of St. Stephen's.'

Will kissed him again then said, 'Do you think there would have been lots of flowers? At my funeral?'

'Not as many as Princess Di got, but yes, quite a few. From all your old gardening fans.'

Will sighed, remembering the events of 1997. 'Christ, those flowers. Remember that sea of bouquets outside Buck House?'

'Oh yes. That was crazy, that whole thing.'

Like the rest of the United Kingdom, and later most of the world, they had followed the royal soap opera from start to very sad finish. They remembered that first television interview given after the engagement.

Charles, looking very uncomfortable, sitting next to his blushing bride-to-be on a sofa. Will had said, 'God, she's so young. What on earth have they got in common?'

'She wants to be a Princess.'

David had thought Lady Diana Spencer had been like one of the Langham girls. Shy, but with a hidden determination to get married, and to a Prince in her case. They had watched the wedding, then the famous trip to America where she had embraced a child suffering from AIDS and admired her for that simple act of kindness and, at the time, bravery.

They had followed her transformation into a style icon, then dotting mother. But as the cracks in the royal relationship began to show, they also followed the drama, the stories leaked to the press and that awful interview after they split up where she had stated that there had always been three people in the marriage.

They had been sad when she had died in Paris in a car accident on the thirty-first of August then stunned by the ensuring mass hysteria of her fans in the weeks afterwards. Once Tony Blair, the new Labour Prime Minister, had crowned her "The People's Princess", the flowers and tributes had started arriving outside the Queen's official residence, Buckingham Palace.

Those first few flowers had given way to an ocean of mourning bouquets, smothering the lawns and streets around the palace.

Will sighed, 'Yes. That was a strange time. And you're right. I don't think I'd have got that many.'

'I'd have picked a few from the garden.'

'Which ones?'

'A few white roses, some of the Sweet Williams… oh, and some of those beautiful hollyhocks you've been growing by the garage. Those are lovely.'

'The red ones? Oh, it would be a shame to pick those. Take the yellow irises from by the pond. Save the hollyhocks.'

David hugged him. 'Okay then. I will. But I'm glad I don't have to, just yet. Night, love, sleep well. And wake up tomorrow, please.'

Will chuckled. 'I'll do my best.'

Chapter Thirty-Four

The day after they had resumed their sexual activity, David was sitting alone in his office, reading his emails. It was mid-October now, and Will had gone off with Cal to buy some spring bulbs to plant in the garden. Over the years, he had added to the already splendid display around the house and the grounds. Apart from the clumps of daffodils that marked the graves of their three dogs, he had planted tulips by the greenhouse, masses of crocuses under the apple trees, and hundreds of narcissi along the walk to the beech wood.

The house was peaceful. Nicky was upstairs cleaning the bathroom. It wasn't one of Ginny's days to work, so David sat reading alone, with Henry the Third at his feet, and Elle Fitzgerald on the stereo. His correspondence was a mixture of fan emails and questions from budding writers, a typical postbag since he had got the website back at the beginning of 1994.

He had been the very first writer in the world to have such a site, and he had been instrumental in H&H becoming the first publishing house to have one too. Both sites had been due to his fascination with CERN, The European Organization for Nuclear Research, based outside Geneva.

When the decision had been made to build the Large Hadron Collider, to study the science behind the creation of the Universe, he had followed its development closely, mainly as a possible plot idea for a novel.

That was because there was wild speculation that it could create a black hole on Earth and destroy the world.

In the end he never wrote such a book, as he concluded that, if there was a black hole, there would be nothing to write about after its formation. However, he asked for and received the CERN newsletter on a regular basis and read about the creation of what would become the World Wide Web, in 1991, by the British scientist, Tim Berners-Lee.

It was for internal use only to begin with, so the scientists could easily share information between their computers, but when it was opened up to the public domain, websites started popping up. By the end of 1993, there were nearly seven hundred of them.

David loved new technology and was what would be referred to in the future as an "early adopter". However, he liked using new stuff, not making it. So even though Berners-Lee published information about programming languages, hypertext and how to make web pages, David didn't feel inspired to learn.

He had never been good at other languages, and even if his books had been translated into over twenty different ones, he could only speak and write English. But he liked the idea of having a website that listed all his books, and being able to write, send and receive messages through his computer. He also guessed right that the idea of the World Wide Web as a way of presenting information would grow dramatically.

He contacted the Computer Department at Bristol University and asked the Director if any of the students knew how to make websites. The Director, who turned out to be a big fan of *The Endless Storm*, had recommended a lad by the name of Harry Pritchard.

He and David had a long telephone conversation about making such a site, and he ended up inviting Harry to come and stay in Fletcher's Cross over Christmas, where he could produce the website, and enjoy some country hospitality. Harry's parents lived in Hong Kong, and he had been faced with a cold, lonely Christmas in his bedsit. He jumped at the chance of good food and making some money from his studies. He had no problem with his hosts being gay.

They collected him, and his NeXT computer, from Stroud station on the fifteenth of December. He was a tall, gangling lad of twenty, red-haired, with a huge Adam's apple, the precursor of computer nerds everywhere. Over dinner that night, they discovered he loved *Doctor Who*, so he and Will spent the rest of the evening going through his collection of tapes, watching their favourite old episodes.

The next day, he started on the website. He took pictures of David's books using a digital camera, uploaded them to his computer, and got started on the layout. David wrote brief background information about each book, extending the back cover notes so make them more attractive to any potential buyer. At that time, Harry could not put in links to bookshops, as they didn't have websites yet.

They worked together up until Christmas, when Harry was totally amazed by the arrival of the best-selling superstar, Charlie Nolan, his wife Cassie, and their children who joined them for the holiday. Charlie looked at the ideas for the website and said, 'Man, that is so cool. I wonder if it would work for musicians too.'

They discussed this over the next few days, but he didn't take it further at that moment.

On December the 31st, 1993, Harry had uploaded the finished website to the World Wide Web, along with an AOL Instant Message address for people to write to David. The author, delighted with the work, paid Harry one thousand pounds in cash for it but then said, 'Now, look. You are one of the few people who know how to do this. In the next couple of years, hundreds of companies are going to want one of these. You should start you own business, making them.'

'For a thousand a pop?'

'God, no. Much more than that. I'm going to write to my publisher and suggest they contact you for such a site. Charge them ten grand. And the next one twenty-five.'

Harry had sat there, in the kitchen, mouth open. 'You're fucking joking?'

'No. You will be in high demand in a few months.'

'I can't afford to start a company and I've got my finals coming up in May.'

'Then wait until after that then start; pick a couple of the best students you know and like, employ them and offer your services through newspaper adds. Call yourself something like UK Website Design Limited, so it is clear what you do. I'll invest enough money for you to get going, for ten percent of the new business. How about that?'

'What about HP Website Design?'

'I think you'll find that HP is already taken...'

'Oh, yeah, you're right...'

Harry had grinned happily and returned to Bristol. David sent an instant message to Gloria Hilton, who also had an AOL Instant Messaging address but no website, giving the web address of "*Connor Lord, Author*". She called from New York that same evening. 'David? What is this?'

'It's called a website, and I've set it up so people can discover my books and read all about them before they go shopping.'

'But they do that by browsing in bookshops.'

'Yes, I know. I do that all the time. But now they can see all of them in one place, read about each one then decide which, if any, or all, they want to buy. In their offices or maybe even at home. It might save them some time and bring in more sales.'

'But hardly anyone has a website or knows how to find one or use one.'

'True. At the moment. But I promise you, Gloria, they are going to be the next big thing. You should get one for H&H. List all your authors, and their books, with pictures of the covers.'

Gloria was silent for a long time. Sitting in her office in New York, she was studying the pages of David's new site. Then she said, 'Shit. That's brilliant. It would be like a permanent advertisement for everyone.'

'Exactly.'

'Who did this one for you?'

The rest was history. David had convinced Gloria Hilton to employ Harry Pritchard to make a site for H&H, and he had set up The British Website Design Company in Bristol that summer, with her as his first major corporate customer. Many more were to follow. When IBM bought BWD in 1998, they paid forty million pounds for it, making Harry a millionaire many times over, and giving David another excellent return on one of his investments. Harry retired after one year with the major American company, hating their culture. He bought a hillside farm in North Wales, shut himself off from the Internet, and started raising sheep there, with his wife and two sons.

*

At eleven, Nicky stuck her head round the study door and asked if he wanted a cup of tea.

'I'll come through and join you.'

Sitting at the kitchen table, he asked, 'How's Theo doing?'

They had never met Nicky's son. Their relationship with her had been different to theirs with Lottie, not for any cultural reasons but mainly age and attitude. Nicky loved to clean, and did a brilliant job, but also liked to go home, never sitting around having a natter at the end of the day.

She had taught David how to cook a couple of her family's traditional Caribbean dishes, to expand his range of global cookery, and his version of her jerk chicken and Trinidad Pelau were pretty good.

However, they weren't close friends as Lottie had been. Also, her father provided the role that David and Will had had with Charlie, so Theo didn't need two more grandfathers.

He was now fourteen and a maths nerd. He also played chess, and was in the county team, based at Stroud Academy, once the old grammar school.

He took the morning bus from Studley Combe and the evening bus home, but if he missed either of those, Nicky had to drive him as the buses only went twice a day now.

'He's got through to the finals. They'll be held in Cambridge and he's so excited as he wants to go there to study Applied mathematics when he's taken his A levels.'

'Has he been in touch with the university?'

'Oh yeah, he's ahead of the game as usual. Booked a meeting with the head of the maths department and everything.'

'He's a clever lad. And how's your mum doing?'

Nana Stanley had started to develop dementia the year before.

'She's okay. Dad's takin' good care of her. He's so devoted, you know. She forgets who he is sometimes, but he's very patient. She can still take care of herself. You know, in the bathroom and such, but she can't cook anymore. Leaves things on the stove and they burn. Dad's taught himself how to cook now he's retired, so that's not a big problem, and I cook too, obviously.'

'Must be hard. We live in fear of developing that. Chances are one of us might. It's an awful illness.'

Nicky nodded, sipping her tea.

'Yeah. It's so unfair too. They have worked all their lives and just when they had a bit of time for themselves, this started. Ah well, life's never really fair, is it?'

David shrugged. His life had been pretty much perfect, but he knew what she meant. Henry, who had followed his master into the kitchen then fallen asleep under the table, now woke up and rested his head on David's thigh, a clear sign he wanted to go out.

'Ah, someone wants a walk. And I need to buy something for dinner. Thanks for the tea, Nicky. I won't be long.'

Fetching the dog's lead and his wallet, he went out through the front door and walked to the butcher's shop to buy something for dinner. It, along with the bakery, were still there, but the greengrocer had gone, replaced by an antique shop, and the post office had been sold off and was now a place that sold scented candles, hand-painted greetings cards, and local pottery.

'Morning, John,' he said to the butcher. 'What's good today?'

'Got some lovely lamb in. Chops or a small joint, Sir David.'

The butcher, John Hoskins, was the son of the original owner, and had taken over the business after his father had retired, just like Todd Young had done at the Fish and Trumpet.

Hoskins insisted on using his title, mainly to impress other customers. As they were alone in the shop, David felt it rather unnecessary.

'A joint please. About three pounds.' He would roast the joint, then make rissoles with any leftovers.

As the butcher sliced through a bigger piece of meat, David looked out of the window at the green. The village looked much the same, but he knew that now ten of the cottages were weekend holiday homes. It was sad, but there was nothing he could do to stop it.

'How's Will doing?'

'Fully recovered, thank God. We had a lucky escape there, John.'

'You did. My dad went bang, just like that, when he 'ad his stroke. But he was very overweight and not healthy like your Will is.'

'True, but we're both getting on a bit now,' he laughed.

The butcher weighed the meat, told him the price, wrapped the lamb in greaseproof paper and slipped it into a plastic bag.

David paid, said "Thank you" and headed out, untying Henry from the lamppost outside.

They walked back together slowly. It was a glorious autumn day, with warm sunshine, so they took their time as the dog sniffed around the verge of the green, weeing when he felt the need to mark his territory.

David thought about Nicky's mother. They were so lucky they still had their faculties. The media was filled with stories of poor care homes and people watching their loved ones fade away in front of their eyes.

By the time Will returned from buying bulbs, David had prepared the joint of lamb, now stuffed with cloves of garlic, peeled the potatoes, chopped up some leeks and made a thick onion gravy. They sat and talked in the kitchen while it all cooked then ate it there, with a nice Chianti.

They were back to normal; both healthy, both happy, both very much still in love.

*

One thing that had dwelt on the author's mind once Will was back to full health had been the casualty nurse's reaction to David when he first went to the hospital.

Once he knew Will was on the mend, he wrote another long opinion piece to the Guardian, outlining what had happened. Back in the days of the main AIDS outbreak, many gay men's boyfriends had had similar treatment from hospital staff, and he wrote, "*nothing seems to have changed since the eighties. It is high time gay couples were given the same rights when it comes to their partners as straight couples have. Ultimately this would be the right to marry, but we could start with some form of civil partnership, allowing couples to be able to support and decide major life choices, just as any straight couple can.*"

The piece generated a lot of mail to the newspaper, and questions once again being raised in the House of Commons. Tory MPs focused on the marriage issue, speaking as if such an idea would bring civilisation to its knees.

The Labour government was more sympathetic, but it took a further four years before legislation was brought in, in the form of the Civil Partnership Act of 2004.

It was passed in November of that year but didn't come into effect until the December 5th, 2005.

On the very next day, David and Will entered into such a partnership at Stroud Registry Office. Ginny, Nicky, Robbie and Charlie were all witnesses.

It had taken nearly fifty years since they had gotten together for the two men to be officially recognised as a couple.

Chapter Thirty-Five

2005 had been a very special year for another reason, as it marked the return of an old friend, *Doctor Who*. Will's favourite television programme had been cancelled back in 1989, due to falling viewer numbers. He had been in mourning for sixteen years when it came back in the capable hands of Russell T. Davies, another huge Who fan. There had been a one-off television film in 1996, made by the Canadian Broadcasting organisation, but it hadn't led to the revival of the series.

Will had sat down in March and been blown away by the new programme. There was a wonderful orchestral version of the classic theme music, amazing production values and a great new doctor, played by Christopher Ecclestone. He also loved the new companion, Rose Tyler, played by Billie Piper. He didn't record it, as he knew the BBC would be releasing a DVD later in the year. David had watched it too, mainly to see Will's reaction. Seeing the sheer joy on his lover's face, he had had an idea for that year's birthday present.

He knew there was a lot of *Doctor Who* memorabilia and themed items on sale, including cupboards shaped like the old police box. He contacted one of the firms that made then, located in Cardiff. Having checked it would fit in the living room, he ordered one. The company agreed to deliver it at twelve-thirty exactly on the big day. Leaving instructions with Nicky to let them in, David took Will out to lunch, saying the meal was his birthday treat.

When they returned from lunch at a nice restaurant in Stroud, Will went into the living room and found a nearly life-sized blue police box, a copy of the Doctor's spaceship, the TARDIS, in the corner of the living room. It was standing opposite the piano, next to the big bookcase. It had shelves in it so he could keep all his video tapes and DVDs in one place, and the light on the top flashed every time he opened the double doors.

Nicky, who had made sure the delivery men had done their job right, had turned to David as Will had stared at his present and said, 'You two guys are like big kids, aren't you?'

'Yep. But look how happy he is.'

Will spent the rest of the day arranging and rearranging his vast collection of tapes and DVDs. The sex they had afterwards was spectacular, despite their advancing years.

Russell T. Davies was already a minor hero in Wisteria House due to another programme he had written called *Queer as Folk*. The BBC series, broadcast in 1999, had followed the lives of three gay friends living in Manchester. It had been the first major series where all of the main characters had been gay.

Although there had been gay characters on television for years, from *Brideshead Revisited* to soap operas like *Eastenders*, an exclusively gay story was ground-breaking.

David loved it as it showed pretty realistically some of the variety of types of gay men; promiscuous, faithful, depressed etc., as well as the gay scene of bars and night clubs along Canal Street in the city. Will had loved it because one of the three was a *Doctor Who* fan, who collected Who memorabilia and had video tapes of the old series in his flat.

When it was announced that Davies would be behind the new *Doctor Who* series, there was some speculation in the media that he would make the Time Lord camp and silly. "Doctor Oooh" was one of the worst headlines in the tabloid press. In the event, it was a triumph, receiving great reviews and guaranteeing the good doctor would keep on fighting evil across the universe for many years to come.

David still enjoyed cookery shows. He watched *Masterchef* avidly, had loved Gary Rhodes and been very disappointed when Jamie Oliver hadn't been nude when his first programme, *The Naked Chef* had gone out in 1999. He loved the young cook's wild enthusiasm, even if some of his techniques made him wince at times.

When Oliver roasted a joint of lamb on the bare bars of his oven, with the fat dripping into a tray below, David had been shocked. 'The mess he's making. Guess he doesn't have to clean the bloody thing afterwards.'

However, he did try it once and they sat, devouring the meat, and agreed it was an amazingly good way to roast a joint. Nicky didn't agree, as David left the oven for her to sort out afterwards.

The one person both occupants of Wisteria House loved on television was David Attenborough. They had started watching him back in the late 1950s, in black and white, as he broadcast his *Zoo Quest* programmes. When the BBC had aired *Life on Earth* back in 1979, David had written to the naturalist afterwards to say how much he'd enjoyed it. Attenborough had written back, saying how much he'd been impressed, and terrified, by *The Endless Storm*. After that, they corresponded from time to time, usually after each new Attenborough nature programme came out, and they once had a very long lunch at the Ivy, discussing the state of the environment and the future of the natural world.

By 2005, David and Will seldom visited London. They were much too old for what was now universally known as the "gay scene", and they had been reducing their time in the capital anyway. They were no longer worried about the threat from Irish bombs, as the Good Friday Agreement in 1998 had ended the threat from the IRA, but other terrorists were targeting London. As they always used public transport, when Islamic bombers attacked three tube trains and a bus on the 7th of July that year, they decided to never go up again.

Like the rest of the country, they had watched the attacks in New York on the World Trade Center in 2001. David had been sitting at his computer when the first plane hit one of the twin towers, and seen a pop-up news item appear at the edge of his screen.

The initial report said a light plane had crashed into the office block. He had called Will in from the greenhouse. They turned on CNN just in time to see the second plane hit then had sat, transfixed, watching the terrible events unfold before their eyes, live.

The next couple of years saw a rise in Islamic attacks, mainly the result of the fiasco that was the Iraq War. When extremists brought their grievances to the streets of London, its citizens, who had relaxed after the Northern Ireland Good Friday Agreement, steeled themselves for another wave of bombs and violence. However, they relented and they did go up to London from time to time, to go to the theatre or for a bit of shopping. They always stayed at Brown's Hotel; David had sold their apartment in Florin's Court for a huge profit, thanks to Agatha Christie's famous detective, Hercules Poirot. When ITV started producing all his stories for television, staring David Suchet, they had him living in Florin's Court, and the block became very well known around the world. He put it on the market and sold it to an American Christie fan. He felt a bit guilty about making so much money considering how many kids were sleeping rough on the streets of London, so he donated one hundred thousand pounds of the profit to the Centrepoint charity for young homeless people in London.

Even if Brown's looked after them very well, they preferred to stay in the village, the Cotswolds, and the West Country. They had stopped going abroad for their holidays too. Wisteria House had everything they needed.

Their home was, once again, the centre of their world. It had the garden.

It had books and DVDs, music, the piano, their dog, and was safe and very comfortable.

'What more do we need, at our age?' the author had said, and Will had nodded, then added, 'Viagra?' He was only joking. In their entire lives together, they never needed help making love. They had a few toys, like the leather waistcoat, the handcuffs, blindfolds and a couple of cock rings, but seldom used them.

Their civil partnership hadn't made much difference to their everyday relationship except by giving them more security in case of an emergency. One morning, David raised the topic of his last will and testament.

'I am older than you, after all, and the chances are I'll pop my clogs before you. You will inherit the house and a shit load of money, more than enough to last you out in extreme comfort, but most of the loot will go to charity.'

Will, who had also made arrangements that what he had would go to David, asked, 'Um, I've never really known how much you are worth. Is it a lot?'

It was one of their after-breakfast chats and they were sitting in the conservatory, as usual.

David had held out his hand and led his partner into the study, sat him down next to the desk, clicked on his iMac and brought up his investment spread sheets. Will squinted at the figures running down the columns, finally settling on the bottom line. 'Jesus fucking Christ. Seriously?'

'Yes.'

'I always knew you had money but I never knew you were this rich. Is it all from the books?'

'The books, the film and television rights, but mainly my Apple Inc investments. Those shares did awfully well.'

'Well? You're like one of those Russian oligarchs.'

'Hardly that. And we don't own a yacht, do we?'

'By the look of all that, we could.'

'I'd hate one of those. So pretentious.'

Will had studied the screen then said, 'Well, I'm glad you're giving it to charity. I wouldn't know what to do with all that.'

It was the first and the last time in their entire relationship that they discussed David's money. Will never mentioned it again, so shocked had he been at his lover's wealth.

2008

Chapter Thirty-Six

Fletcher's Cross, 17th February

They got home from Oxford at lunchtime. Robbie drove them
back, after a night at the Randolph Hotel. David's talk, his second
to the Oxford Union, had run on until eleven, and he had been
exhausted by the time they got to their suite. Robbie was in the
next room just in case they needed help.

David had been invited to speak about writing this time. He had
talked about all his books, his articles and opinion pieces for
various newspapers, covering sixty years of publication. He had
been funny, clear and very well received.

When the President of the Union had asked for questions, nearly
everyone put their hand up. David answered them for nearly three
hours. Finally, the President chose a young Asian student to ask the
last question. She stood up, a little nervous, and asked, 'What are
the three most important things you think a new writer should
consider?'

David nodded. He had been sitting down for all the questions but
now stood up. He looked elegant and distinguished, in a dark blue
open-necked cotton shirt, grey slacks and black boots. He was still
fit, and his hair, now totally silver grey, had been cut by Robbie's
boyfriend, Raj, especially for this occasion. He was now seventy-
seven years old.

Will was sitting in the front row, alongside their helper. David
cleared his throat and, despite the hours of talking, sounded fresh.

'Three things? Right, here we go. First thing is to write something down. I know that might sound a bit daft, but so many people tell me they want to write, but never get started. If you write something down, you have something to look at, to read, to think about, to change, to rewrite, maybe even tear up or delete and start again. But you must write something down before you can be thought of as a writer. You have to make a start.'

He saw lots of heads nodding, and the audience was smiling knowingly. Must be full of wannabe writers, he thought.

'Second thing. Don't write for money. Now I know that might sound ingenuous considering I'm a rich fucker due to my books…'

Big laugh, and a good smattering of applause, with one or two catcalls and jeers; they were not malicious.

'…the point is, you must write what you want to write, not what you think will sell. If it is good, it will sell. If you can get published, that is. I'll come back to that in a moment. If you write what you want to write, the chances are it will be fresher, more alive and more you. It might be something that no one else likes, but it will be yours and yours alone. I mentioned in my talk that publishers still only want to publish what they know will sell, and seldom take risks. But look at two of the best-selling series at the moment. *Harry Potter*, and the *Twilight* series.'

There were groans from the students, professors and other guests. David laughed, 'Bunch of snobs, aren't you? At least I didn't mention *50 shades of Grey*….'

There was general laughter around the room. He went on, 'Ha. Well, my point is, both of those authors wrote what they wanted to write and were rejected over and over again before one got published, and the other went down the self-publication route. They loved their stories and had enough faith in their writing to think others might enjoy them too. And, low and behold, people did. And publishers discovered millions wanted to read those stories. Many people who didn't read much before loved those new stories, those original stories.

They might not be the greatest literature; well, neither is any of mine, but they are stories that entertain and seize the imagination. And in the case of the Potter books, stories that have kick started reading for children in a way no educational initiative has done in this country at any time. So, write your book, not anyone else's.'

He paused, reached for the tumbler of water on the little table between the two leather armchairs on the dais, took a sip, then turned back and continued, 'Finally, third point. If you want to write, write. Don't let anyone put you off. Write because it is the thing you love to do. Write because there is nothing in this life that gives you more pleasure. Write because the words are running around inside your head and need to come out and be put down on paper or into your iPad or computer. Write because you have to and you love doing it. Period. And that, my friends, is all she wrote.'

He turned and sat down, then beamed at the audience, who stood up as one and started clapping. The President of the Oxford Union stood up, called for calm then she said, 'Well, you've anticipated me there. I just want to thank you, Sir David, or Connor Lord, for a brilliant evening and I hope we can all show our appreciation for your words of wisdom.'

The audience rose again and cheered and applauded. David winked at Will, who rolled his eyes, but he was clapping too. It had been a damn good performance.

It had taken them another half hour to get back to the hotel, followed by a quick brandy in the bar before going up to their bedroom.

They had a double bed now, offered without comment by the management, and David fell into it, leaving his clothes all over the floor.

'God, I'm tired. I don't think I can ravish your body tonight. Is that okay?'

Will, settling next to him, head on his shoulder and one arm across his chest, kissed him and replied, 'No problem. You've given me enough pleasure for one night.'

'Really?'

'Oh yes. Especially that comment about gay marriage.'

The first question had not been about writing. A tall, good-looking American student had gotten the first one, and he had said, 'Mr. Lord, are you for gay marriage?'

'Are you proposing?' Big laugh around the room. David had smiled and added, 'Interesting question about my books.'

The audience had laughed again, as he went on, 'Are you in favour?'

'Very much so,' said the student.

251

'Quite right too. My position is this. I have lived with, and loved, and still love, this wonderful man sitting here in the front row, for fifty years. And I still cannot marry him. We cannot go through that simple ceremony to confirm our love in public, in the eyes of God, that any straight couple can do. We are still waiting to do what a drunk man and woman, who might have met five minutes before, can do in a Las Vegas wedding chapel at the drop of a hat. It is a nonsense, it is unfair, unjust and deeply saddening for both of us. That is my position, and I hope it will change one day. Very soon.'

The audience had exploded with clapping and cheers, without a single dissenting voice. Back in bed, David kissed the top of Will's head and said, 'I meant every word. I have always loved you and I still do.'

'I love you too.'

'Good. Can we go to sleep now?'

'Yes, love. Night.'

'Night.'

They had left after breakfast, with Robbie driving their brand-new Volvo XC60 through the busy city traffic, heading for the A40 and home. His arrival as a permanent member of the Wisteria House team had taken place the year before.

After Will had recovered from his stroke he, and David, had had no further health problems but, by 2007 they were both finding life a bit difficult when it came to driving, shopping, carrying things and cooking every day. They had Cal to do the garden and Nicky to clean their home, but it was the other day-to-day activities that were wearing them down.

David's eyesight, when it came to looking at things close up, was getting worse, and when he burnt his hand touching a hot saucepan that he hadn't noticed was a bit too close to the edge of the hob, the situation came to a head.

They were having Robbie and Raj over for dinner when the accident happened. It wasn't a bad burn, and their two guests arrived to find David holding his hand under the cold-water tap, and Will hovering around with some Germolene lotion, ready to smother the little burnt area on the side of his hand. Robbie took charge and made sure the wound was clean and plastered, as Raj helped Will dish up their meal.

It was a casserole with rice, so that wasn't too difficult. Over dinner, the two older men discussed their situation.

David explained, 'I think we need some help, but we're not sure exactly how to sort it out. We'd hate to have someone here full time, like a butler, but to have someone come in and cook sometimes, or do the shopping for us, or drive us, well, that would be great. But it would have to be someone we could really trust. You hear such horror stories of older people being conned or ripped off by helpers and carers.'

Robbie had looked at Raj, who had nodded. He said, 'Um, it's funny you should mention this because we were going to ask you guys a question anyway tonight. The house next door. The one that's just become empty. Have you got new tenants for it?'

'Not yet. It needs a bit of work doing. A new bathroom and complete redecoration. Why do you ask?'

Raj said, 'I've been thinking of starting my own business. My own hair salon. There's a shop available in Studley Combe which would be perfect. Just a shop. Trouble is, our flat goes with my job, so if I leave, we'd be homeless.'

'Well, that's not a problem. Of course, you can have it. But what's that got to do with our problem?' said David, laughing.

'Well, that the other thing. Robbie hates his new employers. They are a bunch of shits, and his workload is simply crazy.'

Rehabilitation services in their area had been outsourced to an American company, which had slashed jobs and wages and increased working hours. The health authority was happy, as it had reduced their costs, but the team were devastated.

Robbie, sitting back with a glass of red wine in his hand, confirmed, 'Several nurses have already left and I am really pissed off by it. I don't have time to treat my patients properly and have to drive all over the county now, not just this area. I used to love my work but this American lot have just ruined it.'

Will sighed. It was not the first time he had heard this sort of thing. 'That's dreadful, Robbie, cos you're so good at what you do. Look how you helped me get back on my feet.'

David was looking at Robbie thoughtfully.

'Would you be interested in helping us instead?'

'Yes. That's what I was thinking. When you said you needed help, it just struck me it might be just the thing. I can cook and drive. I can certainly shop.

If you needed medical help, I'd be on call. If we lived next door, I wouldn't have to live in your home, but would be close enough to help in an emergency.'

David looked at Will, who grinned. 'Sound brilliant. Yes, please, we accept.'

He raised his wine glass and they toasted the new situation. Raj looked at his boyfriend and said, 'Wow, one minute we're weighed down with problems, the next everything is perfect. Oooh. Can we choose the new bathroom?'

David burst out laughing. 'Of course, and you can pick the paint colours too.'

'Yea!'

Robbie rolled his eyes and said, 'God help me. Everything will be pink and gold.'

Raj pouted and replied, 'Not everything…'

*

When Ginny arrived for work the day after they returned from Oxford, she hugged David and said, 'Nice speech, boss. You've got over two hundred thousand hits already.'

She had watched it on YouTube the night before, as the Oxford Union had uploaded the whole thing immediately afterwards.

He smiled and said, 'There were a lot of questions.'

'You handled the Sebastian Rouse one very well, I thought.'

'Thanks. I knew someone would bring him up.'

In 2006, Al Gore, the American ex-Vice-President, had gone on tour with his famous environmental disaster power point, which was later filmed and released as *An Inconvenient Truth*. David and Will had seen the presentation live in London and talked to Gore afterwards. The film had a major impact for a while; Will's only comment after meeting Gore was "It's a shame it wasn't Brad Pitt doing it looking like he did in "Thelma and Louise". Naked from the waist up. That would have attracted even more attention."

Gore was very worthy but dull in person. After the film came out, a Conservative member of parliament, Sebastian Rouse, had made a speech in the House of Commons about the environment and the Chinese in particular.

Rouse was the youngest son of a vastly wealthy English family, who owned a massive stately home in Wiltshire, homes in Nice, New

York and on Bermuda, and a huge yacht moored in Monti Carlo. They were major shareholders in Shell and BP. In his speech, he talked about the threat to the world of one point four billion Chinese becoming tourists, destroying the world as they started visiting it on holiday. He also decried them for building homes using rare woods for their furniture and marble and gold for their decorations, and for buying and driving cars by the million.

He had finished by saying, "Rather than we being forced to change our behaviour, it is a very inconvenient truth that by becoming middle-class, the Chinese will cause more damage to the world than all the oil companies put together."

It was a stupid speech by any standards. David, incensed by it, did his research and wrote an opinion piece for the Guardian titled, "You're a hypocrite, Mr Rouse, and so are we all".

He pointed out that the Rouse family took, on average, five vacation trips every year; two to go skiing in France and Italy, one to their home in the Caribbean, one to France, and one for a nice cruise in the Mediterranean.

Their home was filled with stuffed dead animals; the Rouse family had started, back in the 1800s, the passion among the upper classes for big game hunting, and had a collection of rare specimens, displayed in the Teak Gallery at Rouse Hall, wood panelled with the finest timber from Burma.

They also owned a collection of forty classic cars, and the twelve living Rouses drove eighteen cars between them, including two Aston Martins, a Rolls Royce and a Hummer. He finished by writing,

"I am very lucky, as I live in a beautiful house with twelve acres of land. I can stay home now and spend my remaining days walking about in my own wood, my wonderful garden, enjoying the benefits of fresh air, living in the countryside. I no longer need or want to travel. But I have visited over thirty countries in my lifetime, for work and for pleasure. I have taken many holidays, many flights, many car journeys. For me to start lecturing the Chinese, or any other citizens from developing countries, saying they should not do what I have done, would be abject hypocrisy. If I lived in a high-rise apartment in a crowded city in Mumbai or Chengdu, I would want, and need, a break, be it a run to the coast to sit on a beach or taking a flight to a far-off place of beauty and tranquillity. For us in the West to say no-one can do what we have already done, and keep on doing, is just plain wrong. For you, Mr. Rouse, to demand the Chinese do not do what you and your family do at least five times every year, year in, year

out, and for you to say they cannot have what you and your family have enjoyed for two centuries, is just that; abject hypocrisy."

The ensuing debate lasted for months, as people took sides on social media and raged about the West demanding the East stay put, while they carried on living their best life.

Rouse hit back against David in a piece in the Daily Mail titled, "The Endless Yawn", saying he was too old and feeble to know what was going on. The row could have lasted longer, but the Inland Revenue, looking into the Rouse family fortune, discovered they had been abusing the tax system for years and charged Sebastian, and his father, with tax evasion.

They were found guilty and Rouse had to resign from parliament. One of the Oxford students had raised the topic, asking "Are the Chinese a problem?" David had replied, 'The Chinese government is trying to raise over a billion people out of poverty, which is what any concerned government should do. Have they made mistakes? Yes. Is it going to impact the environment? Yes. But we have learnt in the West what problems our development has caused and we can assist them in not making the same mistakes we made. But to deny them the opportunities we have enjoyed, and still enjoy, without making cutbacks ourselves, is hypocritical. It was when Rouse said it, and it still is."

Ginny grinned and said, 'That was a very good answer from an old and feeble man…'

Chapter Thirty-Seven

Following the dinner when they had offered to employ him, Robbie had started a month later, on a much better salary than the Americans had paid, plus full private medical insurance. The happy couple soon settled into a new routine. One of them would get up around eight, let the dog out, make tea and bring it up to their bedroom. They would then get up, shower and shave and get dressed and come down to breakfast at nine, prepared by Robbie, in the conservatory. Their assistant always let himself in at eight-thirty and got started in the kitchen. Around ten, David would go to his study and work until twelve-thirty, while Will, accompanied by Henry, would go and potter in the greenhouse or walk around the garden. He would natter to Cal, deadhead a few roses or sit by the pond watching the fish.

At twelve-thirty they would meet up again in the conservatory and have a drink, then eat lunch, also cooked by Robbie, at one o clock, then retire to bed for a nap. And have a little fun. They were still active when it came to their sex life, although it tended to be gentler, and a lot slower, than before.

At three-thirty they would come down for tea. Then they would chat with Robbie before either going out for dinner or eating alone at home. David cooked most of the time as before, but Robbie did occasionally, leaving it warming in the oven for them. He did all the shopping.

Sometimes they had a take-out from an excellent Chinese restaurant that had opened in Studley Combe. A fish and chip van still visited the village every Thursday, but now it was run by two hipsters with beards and flame-shaped arm tattoos, who served the cod in recycled cardboard trays, not wrapped in newspaper. The men would chat to Will and David before the "old guys" tottered off home to demolish the food at the kitchen table, once again smothering the chips in vinegar and salt.

They usually went to bed around eleven and then repeated the whole process the next day. When the new bathroom was finished and the cottage redecorated, the two younger men moved in next door. Raj opened his new salon, called, "Hair by Raj", and it was soon doing well. He took on a couple of young hairdressers, both girls, to meet the growing demand.

Nicky still came three days a week and Ginny two. When he wasn't off buying food and the other things they needed, Robbie had a lot of time on his hands. He started reading David's books; he had shamefully told him that he had never read one before and he got totally hooked. Ginny showed him the archive, and he started reading through everything. By the time of the Oxford visit, he was becoming an expert on Connor Lord.

When Ginny decided to retire at the end of the summer, as her husband had already done, and move to a small stone house in a village in the south of France, Robbie added her secretarial duties to his other tasks. David still answered most of his emails himself, but turning away salespeople, filing new articles, those things were performed by Robbie.

*

On the first of June 2008, they celebrated fifty years together. When they had woken that morning, they had just lain together, holding each other, marvelling at their joy at reaching such a milestone. They had a party in the garden, knowing they could all run into the conservatory if it started raining. They had asked the two hipsters with the fish and chip van if they knew anyone who was good at barbequing and they recommended a vast Australian, a huge man with a beard called Marco, who arrived with a massive grill which he set up on the terrace. He marinated pork ribs in beer, slapped huge steaks and Cumberland sausages over the roaring flames and filled the garden with thick black smoke and the aroma of grilled meats. This time they invited just their friends from the village and their families, and had a wonderful, friendly, local party.

*

In 2009, David, who hadn't written a new book by Connor Lord since *Here is the News*, started on his first ever collection of short stories. He read hundreds first, by old masters like Stephen King, Roald Dahl and Somerset Maughan, before making a start. A year later, he had twenty finished, looking at everything in a not-to-distant world, from music to politics and money. He called the collection, *The shape of things undone*, a not-so-subtle tribute to H.G. Wells' *The shape of things to come*. He sent them to Morris, who forwarded them to H&H.

Gloria Hilton had just retired from Hilton House, but her replacement, a very bright man called George Calico, thought them reasonably okay, but he wasn't sure they would sell. Short story collections seldom did, even if written by the grand old man of British publishing. H&H was currently focused on chick lit, self-help books and celebrity autobiographies. More out of respect for Connor Lord than with any real enthusiasm, he wrote back to Morris saying he would publish then, but in very limited numbers. He picked sixteen to publish, rejected one and sent a suggestion for the other three.

Calico knew that, in 2018, it would be fifty years since the publication of *The Endless Storm*. It still sold quarter of a million copies a year and was H&H's best ever seller by a mile. He wanted to produce an anniversary edition, hardback only, with the three short stories he had selected added at the end. This was because they all told stories related to climate change and Connor Lord's classic "end-of-the-world" novel.

Morris relayed this suggestion to David, who was delighted by the idea of an anniversary edition. He was disappointed about Calico's less than enthusiastic reaction to the short stories but accepted the limited run, as long as the rejected story, "*I am what I am*", was included in the collection. Calico reluctantly agreed. It wasn't a bad story, just "silly", he said. He ordered five thousand copies in hardback and a limited paperback run of ten thousand. For an author who usually sold in the millions, this was a bit of a slap in the face. But the wheels were set in motion for the anniversary edition, with cover designs considered but it was all done hush hush, as the date for publication was still nearly seven years away.

David reread the three short stories Calico had chosen, as he wanted to make sure they fitted with his classic novel. He hadn't written them to go with it, but now made sure they would not spoil his old classic.

Back at H&H, they released the new stories, with very little publicity, and the only review was in the Time Literary Supplement, in a general round-up of short story collections. The reviewer was unenthusiastic about *The shape of things undone* and, for the first time, it looked as if a Connor Lord hardback would be remaindered, with unsold copies returning to H&H. The paperback didn't sell well either; most sales were at airports, where people grabbed a book without taking much trouble to read the back cover. Some older people bought it because they recognised Lord's name and were disappointed to discover it wasn't another blockbuster.

Calico, at the next H&H board meeting, was critisied by one director. 'Why on earth did you agreed to publish this shit? The old guy's past it.'

Back in Fletcher's Cross, David was a little saddened by the sales, but as he didn't really need the money, he shrugged, gave Will a kiss and said, 'You can't win 'em all.' Will could tell that he was very disappointed; David still had an ego.

*

Luck had seldom played a part in David and Will's lives thus far, but now something extraordinary happened. A young person, Kale York, was flying from Dallas to San Francisco to take part in that year's Gay Pride March. Kale, self-named and vegan, was in the process of transitioning from a man to a woman, their true form. After years of bullying as a child, a period of self-harm, and endless sessions with therapists, Kale had finally begun the physical transition journey to their true form.

Sitting down in the airport, waiting to board the flight, Kale noticed someone had left a copy of *The Shape of things undone* on the empty chair next to them. Kale picked it up and it opened at the story, *"I am what I am"*.

It wasn't a story as such, but the imagined inaugural speech of America's first non-binary President. Kale read it before boarding, and three times during the flight, with tears streaming down their face.

Someone had finally understood their situation, in beautiful prose. The last line in the new President's speech struck Kale in particular. *"What a joy it is to celebrate the glorious diversity of the minority."*
On reaching San Francisco, and taking the bus in to their hostel, Kale saw a "We print your own T-shirt" shop next to their accommodation. Kale went in and had that last line printed on a T-shirt.

Kale went to the march the next day with some friends and a group of fellow trans individuals, and they all noticed the T-shirt. Kale told about the book and the story. After the march, bookshops around the city were swamped by people asking for a copy. Twitter, finally coming into common use, lit up with images of the book, and that final line was quoted, over and over again.

People got it printed on T-shirts, mugs, mouse mats and one fan made a huge, rainbow coloured poster. The head of fiction at H&H received an email from his sales manager five days later. Amazon were screaming for copies of the collection, as they now had over twenty thousand orders for it. The head of fiction contacted his boss, who contacted Calico. Several phone calls later, a rush printing job for a second edition of fifty thousand paperback copies went through; the third run was for a hundred thousand and the fourth another hundred thousand.

Then Calico was called by a trans activist, offering him five hundred dollars for a hardback copy. He logged onto Amazon and discovered that second-hand hardback copies were being offered for anywhere between four hundred and a thousand dollars.

The shape of things undone hit the top of the NY Times best seller list and stayed there for six months. At the next board meeting, the director who had attacked Calico for publishing "this shit" now demanded to know why he hadn't printed more copies in the first place. Calcio's response cannot be repeated here.

The book was rereviewed in the TLS, this time with glowing praise, and the fiction editor called Robbie, asking for David to give an interview. His response was "Tell him to fuck off..."

David now found himself an icon of the trans movement. It was a very strange experience for him, to have so many young people as fans again. They wrote to him of their experiences, including the derision and bullying they had had at school, in life and on social media, and how much the story had meant to them. He was very touched.

One day he received a letter from Ian McKellen, the actor who had played Gandalf in *The Lord of the Rings* trilogy. They had met several years before, at a gay rally organised by Stonewall in London.

In it he wrote, "*You must be feeling odd. I did. After what seems like centuries of work, to be hailed as a youth icon is very strange, but at least they haven't turned you into an action figure. Yet.*"

David had the letter framed.

George Calico called David personally once the book was a success and apologised for not giving it more publicity to begin with. And for suggesting the "*I am what I am*" story should not be included.

He said, 'I didn't even understand half of the expressions you used, like pansexual and non-binary.'

'That's partly why I wrote it, to inform people or get them to look those words up. People going through transition suffer terrible abuse at times. They need to be treated with more kindness and understanding.'

His publisher sighed down the phone, 'I didn't even know about trans people. I'd heard of transvestites, but not transexuals.'

'You'll hear a lot more about them in the future, I'm sure.'

Calico chuckled and said, 'Once again, Connor Lord is way ahead of the game.' He paused and went on, 'Sir David, I fully understand the irony of the fact that the story I thought was silly has saved this whole thing.'

David laughed and replied, 'Mr Calcio, it was a very silly story, but modern life is very silly, and that's why I knew it would do so well…'

Chapter Thirty-Eight

'Should we be sleeping together tonight? Isn't it unlucky?'
'I'm not sure. That's for a normal bride and groom. As we're both going to be grooms, I'm not sure if it applies to us.'
It was the night of March 28, 2014. At midnight, the Marriage (Same Sex Couples) Act 2013 would come into force, and gay men and gay women would finally be able to marry. Across Britain, some couples were already waiting outside churches, hoping to be the first to get married in specially arranged services. David and Will had chosen to be a bit more civilised and would be tying their knot at two the following afternoon.
The Conservative/Liberal Democrat coalition had finally passed the same sex marriage law after starting formal consultations in 2011. It has passed with support from the Labour Party, with one hundred and sixty-four Conservative members of parliament voting against it.
The Prime Minister, David Cameron, had been a vocal supporter of gay marriage since 2007, often facing a lot of criticism from his own party. But the real hero of the movement, in David Manners' opinion, was a gay rights campaigner called Peter Tatchell.
From his time with OutRage in 1992, Tatchell and his fellow campaigners had put up with physical attacks, verbal abuse and constant ridicule in their efforts to get gay people the same rights as others.
He had reluctantly accepted civil partnerships but called them a second-class solution; he had persuaded Cameron to actually have a Conservative Party policy when it came to gay rights, and forced

Boris Johnson, when he was Mayor of London, to state publicly his support for gay marriage. The Act of 2013 was a tribute to Tatchell's determination to win equality for gay people, and David thought, if anyone deserved a knighthood, it should have been Peter.

Now, he and Will could get married, and in a few hours, they would. St. Stephen's was in the hands of a very charming vicar again, a forty-year-old man with a blond wife and two children, who had returned the church to more traditional services and rebuilt his congregation. It was not up to the levels of the sixties, but David and Will had started going on Sundays again two years before, and their latest Labrador, Henry the Fourth, had learnt to wait for them, tied to the bench outside, just as Flush had done over fifty years earlier.

The Reverend Jack Wilson had been thrilled to be asked to officiate at their nuptials, as he had been an ardent supporter of gay marriage. The fact that he would carrying out the first gay marriage in the Cotswolds had only added to his excitement.

Getting into bed, they hugged and lay together, Will's head on David's shoulder, one hand across his chest, as they had done so many, many times before. David kissed the top of his head, still with a full head of hair, but now it was snowy white.

Will asked, 'Are you sure you don't want to marry a younger guy?'

'I am marrying a younger guy.'

He was eighty-three, and his husband-to-be seventy-nine.

'You know what I mean.'

'There is no-one, alive or dead, I would rather be marrying or spending the rest of my life with.'

'That might not be as long as it would have been if we'd married years ago.'

David laughed. 'I know. Bit of a let-out clause there, isn't it?'

Will tweaked his nipple and chuckled.

'I feel the same about you. Just so you know.'

'Good. Do you want me to ravish your body?'

There was a pause as if Will was seriously considering this offer before he said, 'As much as I would appreciate the effort, perhaps not. We don't want you to miss our big day.'

'At least I'd die happy,' his lover replied, echoing a conversation they had had some years before.

'True, but that would leave me having to tell people why you weren't in church, and I'm far too old to go through something as embarrassing as that.'

That earned him another kiss.

'Well then, we'd better get some sleep. I love you, William Forman. Very much.'

'And I you, Mr. Manners.'

*

They had breakfast in the kitchen the following morning. The house was filled with people as the conservatory was being set up for the wedding reception, with a buffet in a marquee that led from the double doors, over the terrace, onto the lawn. Their guests would be sitting at tall round tables, with stools, dotted around one end, with a dance floor at the other. There would be no place settings, so people could circulate freely. Charlie Nolan's eighteen-year-old son, Ross, who was a DJ, was doing the music. The house was already buzzing with activity.

The wedding planner was there, sorting out the flowers, and the catering team had arrived. Will and David ate quickly, then went for a walk in the garden before going upstairs to change. They would be wearing simple grey suits, white shirts and blue ties. They had seen some pictures of gay weddings from other countries that were ahead of the United Kingdom in allowing gay marriage and had decided that matching white suits, or Edwardian frock coats, or wedding dresses, were not for them.

Will had toyed with the idea that they could both dress as two of the many versions of *Doctor Who*, but David had nipped that idea in the bud very quickly

At one-forty, they walked to the church together, bringing Henry along. He was wearing a red bow tie around his neck and, for once, would be attending the service, not tied to the bench outside. There were cars everywhere. Their two best men, Robbie and Charlie, were waiting for them outside.

'It's packed in there. Everyone's come,' said Robbie.

David snorted, 'I should think so too. They have been waiting long enough for us to do this.'

They went in together, then sat apart, one at the end of each front pew, waiting for the vicar.

Pachabel's Canon was being played by the organist; Will had finally got his wedding music of choice. David looked around. All their old friends and colleagues were there, including Gloria Hilton and Morris Holt. Murrey Holt had passed away three years before, at the grand old age of ninety-nine.

At exactly two o clock, The Reverend Wilson came and stood in front of the altar, and they were off. He followed the traditional wedding service, with no self-written vows, the only change being that he asked them both if they would take the other as their lawful wedded husband. Robbie handed each one their rings.

When the vicar said, 'I now pronounce you husband and husband. You may kiss your husband', they did kiss and the whole congregation stood up and clapped. As they went into the vestry to sign the register, Charlie sang *What a wonderful world*, the classic Louis Armstrong song.

They all walked back to the house in a long procession. The happy couple stood at the front door, greeting their guests, before fetching some food and sitting down at a special table right in the centre of the conservatory, so all their friends could see them.

When everyone had eaten, Charlie tapped on his champagne glass and called for silence for "one of the grooms".

David stood up, beaming with joy. He looked up through the glass roof of the conservatory and said, 'Well, the world hasn't come to an end yet.'

The guests laughed, as they were well aware many right-wing and evangelical people had said it would if gay marriage was allowed. He waited for the noise to die down and went on, 'For those of you who are used to more traditional weddings, today must seem a bit strange. Instead of a beautiful blushing bride and a handsome young groom, you've got two wrinkly old guys, barely able to stand. We have had two best men: one holding the wedding rings, one carrying a defibrillator, just in case. There will be no father of the bride speech, because there is no bride and, even if there was, we'd need a Ouija board to hear from any of our parents, considering we're both around eighty. And even though we have been putting rings on each other over the last few years, this is the first time we've put them on our fingers…'

The gay crowd laughed; it took a bit long for the rest of the guests to get the joke, then there were groans, cheers, clapping and a huge round of laughter.

Will muttered, 'Oh God...' and covered his face with his hands. 'We asked you not to give us any presents, as we have way too much stuff anyway, and, at our ages, to wake up alive each morning is all we need. We wanted you to give money to the Centrepoint charity for homeless young people instead, and I can inform you that you have raised over fifty thousand pounds. We thank you all for your generosity, from the bottoms of both our hearts.'

There was a murmur of self-satisfaction from their guests and a smattering of applause.

David grinned and went on, 'Will and I have waited a long time to get married. To stand in front of our friends and speak of our love for each other and the desire to spend the rest of our lives together. We've done this whole thing arse over tip... not a first for us, of course...'

There was more laughter from the gay contingent.

'... we've spent our lives together and have now got married. It took too long. Much too long. But now we are married. In the eyes of God, and our local congregation, and all of you. Thank you for coming and sharing this moment with us. And enjoy the party.'

Everyone had clapped, then started talking. The volume of noise rose, then the music started. It was *The air that I breathe*. David looked at Will, who shrugged and grinned; he'd asked Ross to play it especially. As they took a turn around the dance floor, David whispered, 'If you fart, I'm filing for divorce.'

Others joined them on the dance floor, another track, a more upbeat one, played and it became a real party. They circulated, chatting to people, catching up with old friends. After a while, David noticed Robbie standing alone, looking a bit down. He and Raj had split up the year before. The hairdresser had dumped him for an eighteen-year-old bicycle courier and moved out of the cottage next door into a flat over his salon in Studley Combe. Robbie had been devastated and was still single. The new groom went over and put his arm around their friend's shoulders.

'Are you okay, Robbie?'

'Not really. This all reminds me of what we could have had. I might never get married now.'

'Hopefully you'll meet someone new.'

'At least you didn't say I would definitely. I hate it when people say that. So many gay men never meet the man of their dreams.'

'Many straight men, and women, too. Will and I have been very lucky.'

'Fifty-six years. It's amazing.'

'And never a cross word. Or another man. Just us.'

'It's beautiful. It really is. Congratulations by the way. I should have said that before.'

'Thank you for being one of the best. Men, that is.'

They laughed. Will came up and slipped his arm around David's waist.

'It's going well.'

'It is, husband of mine.'

'Ha! That's so nice to hear.'

They kissed.

'No regrets?' asked David.

'Only that we had to wait so long…'

Chapter Thirty-Nine

They didn't go away on honeymoon, but just went upstairs when the guests had left, to their bedroom. There was a rose on the bedcover and Robbie had laid out two sets of special bedwear for them; T-shirts with *Groom* printed on them and two pairs of black silk boxer shorts. They were so exhausted they just fell asleep, in each other's arms, oblivious to the caterers clearing away the reception below.

In the morning, Robbie brought them breakfast in bed and suggested they stayed put while the marquee came down.

'Nicky's watching the contents of the house like a hawk. They won't steal anything. And Cal's guarding the conservatory plants and making sure they don't damage the lawn. I've taken Henry for a walk and fed him as well. Everything is under control. Just relax and take it easy.'

He opened the curtains before he left and they sat in bed, eating bacon sandwiches and drinking tea.

'Have we ever had breakfast in bed before?' Will asked.

'Not here. We had it in France once. That time we stayed in Nice. We got croissant flakes everywhere. And I do mean everywhere if you remember. Not sure I like it, to be honest.'

Will laughed at the memory. Yep, they had got everywhere…

'Me neither. But these are good sandwiches. And it was a wonderful wedding.'

'Yes. You looked lovely.'

His husband chuckled,

269

'You too. Very dapper and smart. Oh, and Charlie's speech. I almost cried.'

Charlie had stopped the dancing after about an hour and given his best man's speech. He had told the guests about his life with the two of them.

How he had been born a bastard, who lost his mother and never knew his biological father. Then he had said, 'But I was the luckiest baby on earth. I was raised by my amazing gran, Lottie, and these two guys. My two dads. They taught me everything, from riding a bike to having a wank. No demonstrations, I must add. They have helped me, guided me, advised me and been my rocks, my foundation.'

He had paused and smiled at them then gone on, 'Adult men are never meant to show weakness or be insecure. That is why so many suffer from mental health issues. As we now know. Men are meant to hold down a job, be great providers, perfect husbands, ideal fathers, and protectors of their families no matter what. Most try, many fail, and one reason, I think, is because they don't have anyone to turn to, to talk things through with. Or they feel ashamed if they need to talk in the first place. I said I was a lucky baby; I've been an even luckier man.

Yes, I've had a great career as a singer, and sold millions of albums. Yes, I married the most beautiful and wonderful woman in the world, and yes, I have the three most amazing children any man could ask for. And the best agent in the business…'

Morris Holt, who was happily drunk at this point, shouted out, "Damn right" and almost fell off his stool. The guests laughed as Charlie went on, 'but the thing that has made me the luckiest man on earth has been having these two guys there beside me every day. Not physically nowadays, but they have always been there if I needed to talk, was worried about something or needed guidance and help. Be it about money, or just the stress of living. I could, and still can, always call them, always visit, and they have been there to give me a hug, or advice or encouragement. Never asking for anything in return; a good dad should never do that. I love you guys. You mean the world to me. So, raise your glasses, everyone, to the two best dads a boy could have ever had. David and Will.'

In bed, David said, 'Yes. That was wonderful. He's a great guy is our Charlie.'

'He's our son. Always has been. And you didn't have to sleep with his mother to get him…'

David chuckled, 'You twat. Talk about not forgetting anything. It was a great speech. And oh, Ross did a brilliant job with the music. He's got real talent, that lad.'

'Fit too. Like his dad.'

'Well, with Cassie as his mum and Charlie as his dad, he was bound to look bloody amazing.'

'He is.'

David looked around the room then said, 'I love my wedding present. Very much.'

It was hanging on the wall at the end of their bed. It was a new, original watercolour and ink drawing of a map of Middle-Earth, from Tolkien's *Lord of the Rings*. Painted by a local artist who lived just outside Bath, called Petunia Crow. She had got permission from Christopher Tolkien, the guardian of Tolkien's legacy, to do a few of these; it looked ancient, as she had aged the parchment by staining the paper with tea.

As David had told Melvyn Bragg back in 1980, he had always been a fan of the three books that made up the saga. Will had never read them; they had only had the bible at home when he lived with his mother, and although he had seen a copy on the bookshelves in Wisteria House, he had always been intimidated by the sheer size of it to get started.

When Peter Jackson's first film came out in 2001, they had gone to see it in Stroud. On the way, David had asked if Will wanted to know a bit about it before they saw it. Will had shaken his head and said, 'Best not. If you don't like it, or it misses out something you loved in the book, I don't want to know. I'll take it as I find it.'

They had both been blown away by *The fellowship of the ring*. Yes, it wasn't exactly like the book, but it was damn close, as David said afterwards. Will had just been stunned by the whole thing, and shocked when Gandalf had died. His partner had held off telling him what happened next.

They had seen *The Two Towers* in Stroud again, but Morris Holt had got them tickets to the London premier of *The return of the King* in 2003, and they saw it in Leicester Square. They had had to stay in their seats when the film ended, until Will stopped crying.

When Aragorn had said, "My friends, you bow to no-one" and had knelt, along with the whole cast, in front of Frodo and the other hobbits, he had burst into tears and kept on crying to the very end. Needless to say, Will had bought the boxed set of all three films (extended versions) and occasionally binged watched them in one day, with David bring him a supply of meals and drinks as he relived the epic story.

David's wedding present to Will was in the wardrobe; one pair of Lobb's boots, made by the same apprentice who had made his original ones back in 1958. This time they had visited the man, Simon Lovell, and he had made proper lasts of Will's feet. He was the current Master Last maker. He had been fifteen when he made the first pair of boots, and was about to retire, aged seventy-one. He had said, 'That tenner you sent me, sir. Paid my rent for three months. Absolute life saver it was, cos they paid me very little back then.'

Will had said, 'Those boots...I still wear them today sometimes.'

'As you should, sir. They are meant to last a lifetime.'

David put his plate on the bedside table and lay back against the pillows.

'Hmm, yes, it was a very good wedding.'

'It was. And that was a very good bacon sandwich...'

David put his arm around Will's shoulder, hugged him and kissed him on the cheek and said, 'So, husband, how would you like the spend the first day of our married life?'

Will looked up at the ceiling then said, 'Well, husband, I hate to mention it, but we have not actually consummated this marriage yet.'

David looked at him in mock horror and replied, 'Good Lord. You are absolutely right. Well, we'd better do that before we do anything else...'

2018

Chapter Forty

Fletcher's Cross, 3ʳᵈ July

'I'm not going to do any more interviews. She was checking her emails all the time she was asking me questions. A bit rude if you ask me.'

David had just finished a Facetime chat with the Guardian's Book Section editor about the anniversary edition of *The Endless Storm*. He snapped the top down on his MacBook Pro, took off his reading glasses and looked at Will, opposite him across the teak table in the garden. They were sitting on the lawn, enjoying the sunshine. Henry was laying at their feet, sleeping peacefully.

'It's about time. Is there anything more left to be said?'

'No. You're absolutely right, love. I've said far too much for far too long. I shall remain silent from now on as far as the outside world is concerned.'

'But you'll still talk to me?'

'I shall. Always. With pleasure.'

'Good. I love talking to you.'

Robbie joined them. He was carrying a tray with a jug of Pimm's, three tumblers and a bowl of crisps. It was midday on a sunny summer's day, and a gloriously fine one at that.

'Thought you two guys might like a Pimm's'

'You've bought three glasses,' said Will.

'Have I? Oh, well, I'd better have one with you then.'

He poured their drinks, handed them over, put the crisps on top of the computer so they both could reach them and sat down in the third chair.

'How did it go?'

'She was multi-tasking. Bit rude really. I don't want to do any more interviews. If anyone asks, tell them to bugger off.'

'Good. Okay. No problem. You don't need to do any more anyway. The book's a huge success.'

The special edition had been published in March that year. Calico had decided to go all out and run a major pre-publication marketing campaign, then ordered a print run of one hundred thousand hardback copies. They had sold out in three days. A reprint had also sold out, and the paperback edition, (Calico had changed his mind about having one due to the high sales of the hardback) which had been released on the first of July, was also selling well. Most reviewers felt the three new short stories published at the end were excellent additions to the book. Sir David Attenborough had written a preface, saying how much he had loved the original book, and how much it had influenced his subsequent thoughts on the future of the environment. He said he loved the new stories too.

The first, *Karma's a bitch*, was about a group of preppers in the United States. They had built a community of underground shelters to retreat to, come the apocalypse, under the shadow of a dam in Texas. They were all like-minded souls, who hated the federal government, and refused to pay federal taxes. David painted a picture of a state in revolt, with no money available for public finances, including dam maintenance.

When the preppers finally had to retreat to their underground shelters, they locked themselves in with their food and their guns, to wait until "it" was all over. Unfortunately for them, the dam, badly looked after and in desperate need of repairs, broke under pressure from extreme rainfall and hurricane force winds, and their homes were covered by a new lake. The story ends as the first of the shelters starts leaking.

The second story, *Darwinism*, told the story of an obscenely rich billionaire who had built a luxurious shelter on the top of a cliff in Hawaii, surrounded by a high wall, and filled with fine wines, food, art works and luxury goods for his second wife, three daughters and two mistresses. The building had a cinema, an indoor swimming pool, twelve bedrooms, fifteen bathrooms and a vast living area.

The project had been protected by a group of former mercenaries, all well-trained and extremely fit young men, who had kept the locals away during construction. As the long winter caused by climate change swept over the islands and the temperature dropped below minus twenty degrees C, the billionaire and his entourage had flown in on helicopters.

The compound had been surrounded by locals, desperately seeking shelter from the cold. He had ordered the gates closed, the high voltage wire running around the top of the wall switched on, the doors locked and the shutters over the windows brought down. He then asked for the head of his security team to see him in his study. He ordered him to have six men patrol the outside of the house twenty-four hours a day, in three shifts, despite the intense cold. That was when things went wrong.

He had treated his team very badly, with low wages, poor food and cheap beer for months, knowing he could replace them at any time. He could, before the shutters went down, but now? The head of security had smiled and simply shot the billionaire through the heart; his men had wrapped the body in one of the late man's expensive Ziegler Mahal rugs and tossed him off the cliff before they settled down to enjoy his home, and his women, as they waited for the sun to shine again.

David was far too subtle to mention anything about the survival of the fittest.

The final story, *A very silent spring*, was a sad tribute to Rachel Carlson's book. It described spring coming to a future English countryside, post climate crisis. No snowdrops or crocuses appearing in January, no daffodils bobbing in the breeze, no lambs bleating in the field, no frogs croaking in ponds. No blackbirds singing at the dead of night. Just frozen ground, dead trees, grey skies. It was beautifully written but very depressing, and an awful warning of potentially what was yet to come.

David hadn't written anything new since their wedding. He and Will had spent the last four years simply living and enjoying their own company. Most days they would go for a walk if the weather was good. That summer, they had spent many fine days in the beech wood. They would take a canvas satchel with coffee, sandwiches and some cake and make their way through the heather garden, through the rhododendron walk and sit down on the beech trunk.

There they would eat and drink, watching Henry nose around the bracken, looking for sticks. He always found one and would carry it back to the house in triumph.

The recent bout of interest in the ancient author had disturbed this peaceful routine.

Robbie, sitting back and sipping the cold Pimm's, asked, 'What do you boys want for dinner tonight?'

'How about we take you to the Frog? Save you cooking?'

'I have no problem with that. Lovely. Do you want me to book a table?'

'Yes. Please. It will give you a chance to chat to Brad.'

Robbie had started dating the new chef who had taken over the old restaurant. The original owner had lost interest, and a succession of bad chefs had damaged its reputation. It had lost customers and looked as if it was going to close down for a while. Brad Simmons, a very dynamic new chef, had bought it, updated the menu, revitalised the interior and now it was back as the premier destination eatery in the area. Robbie pulled out his phone and called his boyfriend, who confirmed they could have a table for three at six, their preferred time for eating.

'Ask him what the specials are tonight.'

Robbie listened and said, 'Fresh crab cakes. And he's got some Dover soles too.' David grinned. He loved sole.

'Can he save me one. Us. One each in fact,' as Will pointed at himself.

Robbie relayed this order then laughed and disconnected. 'He says, at six in the evening, they should have two left. We'll be the first ones in.'

They were. Other guests started arriving from six-thirty onwards. The three men were just starting on their crisp crab cakes when Robbie muttered, 'Oh fuck...'

Raj had come in, with a very young skinhead lad, who had diamond studs in each ear lobe. He looked sulky and annoyed and they clearly heard him say from across the room, 'Christ, this place is like an old people's home.'

Raj, who was looking much older and fatter than the last time they had seen him, shooshed the lad. The waiter gave them menus then asked if they wanted anything to drink.

'Lager,' said the skinhead.

'Oh, I'm sorry, sir, we don't sell lager. We have some fine real ales.'

The lad stood up and shouted, 'I don't want any fucking real ale. I want a fucking lager. What a dump this place is,' and stormed out. Raj gasped, apologised to the waiter and followed him. They didn't return.

The other diners had watched the row in silence, but no one commented. They were far too well-bred to do that. Now they started chatting again over their meals. Robbie had watched this little drama with his mouth open. He turned to his friends and said, 'Wow. That must be the fourth

or fifth guy he's been with since we broke up. Maybe many more than that. What a little shit.'

'His choice,' said Will.

'Yeah.'

A few minutes later, Chef Brad came out from the kitchen, greeted a couple of other guests then came over and gave Robbie a kiss on the lips.

He was very fit, around thirty-five, with cropped black hair and blue eyes.

'Hi guys. Sorry about the disturbance.'

'No worries. We love a floor show,' said Will.

'I don't. He's banned from now on. Both of them are.'

David laughed. 'Not sure the skinhead would ever want to come back anyway.'

Brad nodded. 'Raj always brings his new ones here to show off how much cash he's got.'

Robbie winced and Brad patted his shoulder. 'You had a lucky break there. You might not have felt it at the time but you did. And now you've got me instead.'

'I know. Thanks.'

'No thanks required. Never thought I'd meet a hot guy out here in the countryside.'

David laughed, 'Me neither,' and winked at Will.

'Gotta go back in. Hope you enjoy your meal.'

'Thanks, Brad,' said David, 'The crab was delicious.'

The chef nodded and walked back to his domain. The rest of the meal passed without drama; the soles were stunningly good, and they took a taxi home after coffee. They had come by one too, so Robbie could have a drink. At the front door of Wisteria House, they went their separate ways, Robbie to his cottage to wait for Brad to come home after the restaurant closed, and David and Will to go in and let Henry out.

They stood, as they had done so many times before, on the terrace, looking up at the night sky, as they waited for their current dog to do his business. In bed, spooning, David stroked Will's furry belly.

'Am I getting too fat for you?'

'Not at all. I like it. Makes you even more cuddly than before.'

His husband sighed happily. 'How is it you know exactly what to say to make me feel secure and loved?'

'Practice.'

Will felt something hard growing behind his back, pressing against his bottom. 'Are you considering ravishing me?'

'I was thinking about it. Are you interested?'

'Davey. I've been interested in you ravishing me since 1958...'

Chapter Forty-One

Even though David had not written anything new since his marriage, he had stayed very interested in what was going on in the world around them. He had watched the destruction of the rainforests in Brazil and Indonesia for palm oil trees and soya bean production with misery. The appearance of ISIS, and their terrible acts of violence, often filmed in ghastly videos, had shaken him to the core. Ongoing wars in Africa, and Russia's annexation of the Crimea, and the dreadful Brexit campaign, that took Britain out of the European Union, caused him much sadness as well. He believed in European unity, even if the EU itself did need reform, so he had voted against Brexit, hoping the country could stay in and make those necessary changes. Alas, Boris Johnson's decision to back it, led to the referendum being won by the flag-waving, "Good old England" brigade, most of whom regretted it by the time it actually happened.

But it was the rise of the climate-change deniers that truly surprised and horrified him. Just at the moment that the effects of climate change were beginning to have a real, and very negative, impact on the world, the number of voices saying climate change was a myth concerned him deeply. Attacks were made on experts as their dire warnings of Man's impact on the environment were dismissed as "fake news".

David had long ago stopped watching "Fox News", as he couldn't stand the dreadful people who, even more than before, presented their views and opinions as real facts.

When Donald Trump had announced his intention to run for President, David had thought it a joke, meant to promote the TV personality and his other businesses, not a serious run for the top job in American politics. His supporters on Fox helped him win, and Will and David had watched his face on the big night. David had sighed and said, 'Look at him. He's as shocked as we are.' The next day, he received, by special delivery, a package from Morris Holt. It was a very fine bottle of Hennessy X.O brandy in a box. There was a note included saying, "Thought you might need this. It must be awful being right all the time!" Morris was correct. The next couple of years would have made a saint start drinking. When David saw a picture of Trump looking directly up at the solar eclipse in 2017, he sighed and said, 'I guess it is too much to hope that it blinds the stupid fucker,' and turned off the news.

He was asked to comment many times on climate change, but simply had Robbie reply with a standard message, "Connor Lord deeply regrets that he is too depressed by the climate change denial movement to say anything. It is over ten years since Al Gore's *An Inconvenient truth* added to the message he wanted to get across in *The Endless Storm*, and still nothing gets done."

He had remembered the ex-Vice-President's presentation and film in 2006, looking at the issues he himself had raised back in 1968. As a popular way of getting the message across, it was very effective, but the political chasm that had opened up in American politics, and the growing distain of "experts", meant it had quickly faded away from public consciousness.

There were brighter moments. Fish had returned to the Thames, and the barrier that had finally opened in 1984 worked perfectly, even though there were some suspicions that, in the event of a major spring flood surge up the estuary, it might not be big enough for the job. In Britain, beavers had been reintroduced, the raptor population had risen, and the ecological planting that Will had promoted for so long was becoming more acceptable, and more widespread. Several butterfly species increased in numbers, the swallows still came every year to nest in the eaves of the garage, and their bird boxes were always full.

But the traffic had gotten worse on the roads, with more and more cars and lorries choking the small country lanes around them.

They might not run through the village, but the dual carriageway was very slow to use now, as both carriageways were always busy. In Fletcher's Cross, in the four years after their wedding, twelve more cottages had been sold off to outsiders. Not only to weekenders now, but also investors who bought them to use as AirBnB properties, making more money in a week from letting them out than most local people paid in rent for a month, or two. The result of this was that the young people who had grown up in the village had no chance at all, unless helped by their parents, to buy a home in the area. The prices were far too high. If they wanted to stay, that is. There was little work there, as the farms had mechanised and unless they could work from home, they had to leave to build careers and lives elsewhere.

David and Will watched all these changes take place and listened to the worries and concerns of the locals when they went to the Fish and Trumpet once a week. Even there, changes were apparent. Due to the higher prices of alcohol, it was cheaper for people to buy their booze from a supermarket and drink at home. But some locals still drank there, bemoaning the problems of modern society.

But for the two of them, life was much the same. They loved their home; they loved their garden. They loved their dog, and they loved each other.

For their sixtieth anniversary, they asked Robbie to make them a picnic and take it up to the beech wood then leave them alone. They didn't want the fuss of a party, and he was more than happy to help them have a romantic time together, on their own. As they sat there eating their lunch, David said, 'We're so lucky. We always have been. No hassles, no arguments, no drama.'

'Yes, you're right. We've never had to go through what Robbie's been through, have we?'

Their friend and assistant had, after Raj had left him, experienced a period of mourning, then tried to date again. He had used Gaydar and the new app on his phone, Grindr. They had followed all his trials and tribulations as he had tried to meet new men. When he first showed them the dating app, Will had said, 'It's like the Chinese menu. You can pick one of those, one of those and one of those.'

Robbie had grunted and said, 'Not really like that, these have to pick you back…'

David chuckled, 'No, thank God. I'd have hated that.'

'Me too. All those false starts, and failed hopes. I'm glad he's happy now.'

'As we still are.'

They sat there for a while then David sighed, 'We've been so damn lucky. Living here, all these years. We've very privileged. All that pain and suffering in the world, and we've just been here, in peace.'

'It's all down to your books.'

'No. It's you too. It wouldn't have been the same without you.'

He took Will's hand and kissed it.

'It's you too. I love you, William Forman.'

'And I you, Mr. Manners.' And that wasn't fake news.

Chapter Forty-Two

'Why do they keep changing the names of things?'
Davey looked up, 'I'm sorry? What did you say?'
'Why do they keep changing the names of things? For example, what would you call this?'
Will held up his iPad, which showed a photograph of a fountain on the *Gardeners' World* website.
'That's a fountain.'
'That's what I thought but apparently it's a KWI.'
'A what?'
'A KWI. A kinetic water installation.'
'Looks just like a fountain to me.'
'I know.'
'GW hasn't been the same since you left.'
'G what?'
'GW. *Gardeners' World*.'
Will snorted with laughter and went back to his reading. They were finishing breakfast the morning after their dinner at the Frog. Robbie had made them poached eggs on toast, which they had eaten with real pleasure, and they were now on their second cup of coffee, finishing off some toast with Wisteria House honey, catching up on the news.
Will was reading an article on the *Gardeners' World* website about oxygenating plants for ornamental ponds, except they were no longer called ponds but "water features".
David put down his own iPad, on which he had been reading the

Guardian On-line's letters page, and asked, 'What shall we do today?'

Will replied, 'How about a walk to the beech wood and back. We haven't been there for a couple of days.'

'Excellent idea.'

Henry heard the word "walk" and jumped to his feet from under the table. Will rubbed his head and said, 'Wait for it, Henry. I need to have a word with Cal first, about these lemon trees. They need a bit of fertiliser.'

Cal still looked after the garden, now aided by his son. He and Nell had had two more children, another girl and finally a boy, Adam. The two of them were currently trimming the beech hedges behind the herbaceous borders using the latest electric trimmer, and not doing it by hand as Will had done for so many years.

Will stood up, and came round to where the author was sitting, bent down and whispered, 'You were magnificent last night, you beast.'

David laughed, 'And you, dear husband, were as tight as the first time I had you.'

Will sighed dramatically and replied, 'You wish. Won't be long.'

He gave him a kiss on the lips as Henry rushed out barking onto the grass. Will followed him out of the conservatory, stood at the edge of the terrace and stopped. He gasped, froze then fell headfirst onto the lawn.

David saw him fall and screamed out, 'Will! Robbie! Robbie! Will's fallen down.'

David got up, reaching for his phone. Robbie dashed through from the kitchen and ran to help Will. Cal, who had seen him fall, sprinted over as well. Robbie, turning the fallen man over, shouted, 'Call the ambulance. He's had another stroke.'

David did. It took fifteen minutes for it to arrive, while Robbie performed first aid and kept Will breathing. David knelt next to him, stroking his head, saying 'Come on, Will, come on.'

The ambulance crew intubated him, put him on a stretcher and headed off, as Robbie fetched the car and drove David to the hospital. They sat in the casualty department for nearly two hours, waiting to hear about his condition.

Finally, a young doctor approached them and said, 'Are you here for William Forman?'

'Yes. I'm his husband.'

'Please come this way. We've moved him to a side room.'
Robbie supported David as they followed the doctor. Will was in bed, on a respirator, once again connected to various pieces of medical equipment. He looked very peaceful.

The doctor said, 'I'm very sorry. He's in an extremely bad way. It was a massive stroke. We're keeping him alive with the machines but I have to tell you that we can detect no brain activity at all. We're keeping his heart going, as his body cannot do that without help. With a case like this, the chances of him recovering are very, very remote. Even if he lives, he would be in a vegetative state for the rest of his life. There's been too much damage to the brain.'

'Oh God,' said David, and sat down on the chair by the bed. He reached for Will's hand and held it; there was no response.

Robbie asked, 'No chance at all?'

The young doctor shook his head. 'I'm so sorry.' He looked at David and said, 'As his husband and next of kin, it is your decision as to what happens next, and when.'

'You mean I have to decide when you turn him off?'

'I'm afraid so. Have you been together long?'

'Sixty years.'

The young doctor smiled sadly. 'Oh, how wonderful. Well, he is not in pain. I can assure you of that. I'll leave you for a while, to let you think about things. Once again, I am so sorry.'

He left the room. David squeezed Will's hand again and whispered, 'Oh, Will.'

They stayed like that for an hour, David in the chair, Robbie leaning against the wall. They didn't speak and David didn't cry. Then he stood up, lent forward and kissed Will's cheek.

'Can you go and find the doctor? Will always said he liked growing vegetables but never wanted to be one. After the first stroke. I can't leave him in limbo like this.'

'Are you sure?'

'Yes, Robbie. Please find him.'

He left the room and returned with the young doctor. David nodded his consent so the young man carefully removed the oxygen tube from Will's throat and turned off the pace making machine that was keeping his heart going. As David held one hand, the doctor held the other, his fingers feeling for a pulse. After two minutes he said, 'He's gone, Sir David.'

He had found out who his patient's husband was during his absence from the room.

'Thank you. For being so straight forward and honest. Always best in this sort of situation. What happens next?'

'We will contact an undertaker to collect Mr Forman's body.'

'We have made arrangements with Johnson's. If there are any bits of him that might be useful for someone else, you have my blessing.'

'Thank you. And we will contact Johnson's. Goodbye, Sir David.'

He left the room. Robbie said, 'You have planned the funeral?'

'Just which company to use and what sort of coffins and service we wanted. Just in case we both died at the same time in an accident.'

'I didn't know that.'

David sighed. 'When many of our friends died during the AIDS crisis, we found out most had never even made wills, let alone thought about a funeral. They were too young to even consider such things, most of them. No one knew what their wishes were as far as funerals were concerned. We didn't want that confusion. Can we go home now?'

'Of course.'

David patted Will's hand for the last time. Robbie put his arm around the old man's shoulders as they left the room and, after David had signed the organ donor release forms at the desk, led him to the car park. Once he was settled into the passenger seat, he turned to Robbie and said, 'Can you stop on the way home and buy me a packet of cigarettes? Benson and Hedges Gold.'

'Really?'

'Yes, please. What harm can they do now? And a lighter.'

Back at the house, David walked slowly to the seat by the pond, asking to be alone for a while. Cal saw them come back and looked at Robbie, who shook his head. Cal frowned, muttered "bugger" under his breath, wiped his eyes and went back into the greenhouse. Seeing David was settled on the seat, Robbie went inside and told Nicky what had happened. She sat and cried at the kitchen table, and he joined her.

David opened the packet of Benson and Hedges and lit one. It made him cough and cough, but he kept smoking. Then lit a second. That one was a bit smoother. Henry, glad he was back, joined him, standing at the edge of the pond, watching the fish.

'Don't go in, Flush, old boy.'

The dog looked at him, slightly confused, wagging his tail, then came and curled up at his feet and fell asleep.

David looked around the garden. The herbaceous borders were in full bloom, and butterflies and bees flew all around the flowers. In the pond a single frog sat on one of the lily pads, as tadpoles nuzzled at the edges under the flagstones. A big blue dragonfly flew down and settled on an iris stem, and one of the goldfish came to the surface, gulped some air then dived back down into the depths of the water. The fountain was running and David closed his eyes tightly, holding back the tears. What a stupid last conversation to have had, about what to call a fountain. He watched the droplets fall onto the surface; it was beautiful, whatever it was called. It was a wonderful garden, he thought. A real tribute to Will's skills and passion. He had made it a paradise for them to share for so many, many years.

David looked at the other end of the bench, where Will used to sit. For the very first time in sixty years, he felt overwhelmed by loneliness.

Robbie brought him out a cup of tea an hour later. The stone ashtray by the seat, unused for so many years, was now full of cigarette stubs. He sat next to the author and said, 'Are you okay?'

'Oh yes. I think so. We always knew this could happen. But I can't go on without him, Robbie.'

He studied the old man, who was very calm and seemingly relaxed. He realised David hadn't cried once since Will had collapsed, not by his bedside nor, by the look of it, out here in the garden. Concerned, Robbie asked, 'What are you saying?'

David, who had been staring out over the garden, looked at him directly and smiled. 'Oh, don't worry. I don't mean I'm going to kill myself. Nothing so dramatic. But my heart broke today. I felt it go. I can't live without Will here beside me.'

'You're still a very fit man.'

'Doesn't matter. I can't go on without him. Not after all these years together.'

He lit another cigarette and inhaled deeply. 'I'd forgotten how much I enjoyed these things.'

'I never knew you smoked.'

'We both did. We would sit right here, by the pond, smoking, Will and I.

When he first came to Wisteria House as the gardener. He was so shy. Took ages to get him to loosen up and talk. But he did, eventually. Just sitting here, the two of us, chatting about this garden. We stopped smoking when I wrote that book about tobacco.'

'Not surprised. Are you hungry?'

'Not really. I'll just have a sandwich later, then go to bed. I'll pop into my study for a while and write for a bit when I have finished this ciggie. Can you bring it to me there? Just ham and cheese.'

'Of course.'

'And would you mind sleeping here tonight, to take care of Flush? Let him out for a wee last thing and feed him, of course?'

'Flush?'

'Oh, sorry, Henry. I meant Henry.'

'Of course.'

'Good. You're a great friend, Robbie. Thank you. And thank you for the tea.'

He was being dismissed. He got up and left David, sitting by the pond, in Will's beautiful garden, smoking his last cigarette.

The End

Afterword

David Manners passed away in his sleep that night, on the morning of the fifth of July 2018, less than twelve hours after Will Forman. I found him when I went to wake him up when he didn't come down for breakfast; he looked peaceful, with his eyes closed and a smile on his face. The doctor said his heart had just stopped; I think he had lost his only reason for keeping it beating.

The final thing David did before going to bed that night was complete the last lines, covering Will's death, in the manuscript you have just read. It is the closest he ever came to writing an autobiography; he never usually wrote from a first-person perspective, apart from when he produced Will's gardening books. He left a note requesting that this manuscript be withheld from publication for ten years after his death and expressing his childish delight that he had managed to write their story in forty-two chapters. Douglas Adams would have approved, hopefully.

David left their home on the green, and his vast archive of material, to *The Wisteria House Trust*, along with enough money for them to be looked after for many years to come; his shares in Apple mean it is one of the best funded charities in the country. He also suggested I be made curator, and I have served *The Connor Lord Collection* in that role for the last decade.

David left instructions for the trust's considerable fortune to be used, not just for the house, but for charities helping young people; the homeless, the sexually abused and confused, as well as young writers. Charlie Nolan, the Trust's chairperson, makes sure it is used for those purposes.

I think David died at the right time. His prophetic work, *The Endless Storm*, is beginning to come true with a vengeance, and he would have been deeply saddened that more has not been done to prevent the loss of life, and damage to our homes and cities, that climate change has caused, and will carry on causing, in the years ahead. He hated being right, be it about the weather or Margaret Thatcher.

Wisteria House still gets thousands of visitors each year. Most come in the summer, to enjoy Will's beautiful garden. These have been maintained through the Trust and are a training ground for young gardeners from all over the country, paid for by a separate charitable fund, *The Will Forman Young Gardeners Trust*. The house is closed for three months during the winter, as we host writers' retreats for young authors between November and the end of January each year.

David is buried with Will in St. Stephen's churchyard, and their grave is alongside those of Lottie and Flora Nolan. We place flowers from the garden on them every year on the 5th of July to mark their passing, and as a tribute to a love affair that lasted over sixty years.

Robert Harrison,
Curator, "The Connor Lord Collection"
Wisteria House, Fletcher's Cross, Gloucestershire

July 2028

Author's Note

I do hope you have enjoyed reading this story. If you did, you might enjoy my Gemini & Flowers mystery series. This is a series of ten books, set in the Dorset countryside, where murders and mysteries occur around the village of Rilton Castle and the Ash family. Once again, there are many strong gay characters at the very heart of each story. The same could be said for 'Cow Boy' another English gay romance, set in the Lake District.

I have written 'Nice People', a novel set in the Old Town of Stockholm, and the 'Tall Timber' Trilogy, three books about a young gay porn star, overcoming his past and building a new life in northern California. There is also 'The King's House', a love story set in modern-day Oxford. Other books include a collection of short stories called 'Accidental Murderer', a 'future history' science fiction novel called 'The Stone Age', two ghost stories and several other crime books, including 'Who Dun It?' a classic country house murder investigation by a soon-to-be retiring police officer.

If you want more information about any of my other books, please visit my homepage: http://thegeminiandflowersmysteries.com

If you have any comments about "Life Story" or wish to get in touch anyway, I am always happy to hear from readers. Feel free to drop me an email at

jonathangregorysweden@gmail.com

Jonathan Gregory,
Stockholm, Sweden, May 2022

YC
COMI

GW00458206

GEMINI 2023
PERSONAL
HOROSCOPE

Monthly Astrological Prediction Forecast
Readings of Every Zodiac Astrology Sun Star
Signs- Love, Romance, Money, Finances, Career,
Health, Travel, Spirituality.

Iris Quinn

Alpha Zuriel Publishing

We are born at a specific time and place, and, like vintage years of wine, we have the characteristics of the year and season in which we are born. Astrology claims nothing more.
— CARL JUNG

CONTENTS

GEMINI PROFILE

Constellation: Gemini

Zodiac symbol: Twins

Date: May 20 – June 20

Zodiac element: Air

Zodiac quality: Mutable

Greatest Compatibility: Aquarius and Sagittarius

Sign ruler: Mercury

Day: Wednesday

Color: Yellow/Gold and Light Green

Birthstone: Pearl

GEMINI TRAITS

- Charismatic
- Humor is used as a crutch.
- Could converse with a brick wall
- Uses Arguments to flirt
- Knows a little bit about everything.

PERSONALITY OF GEMINI

Geminis are extremely intelligent and learn quickly. They are astute, analytical, and frequently amusing. They have a childish curiosity and are continually asking new questions.

Geminis have an incredible capacity to judge a person's personality in a couple of seconds, even if they have only met them. They'll be the first to recognize whether someone is bluffing. They are excellent speakers, as well as attentive and sensitive listeners.

Geminis are adaptable, at ease as introverts and extroverts. They respond quickly to the energy of a room. They might be the life of the party or a total bore. They understand how to bring disparate people together and make them get along.

WEAKNESSES OF GEMINI

Geminis dislike being alone. This is why they have wide social circles and are always accompanied. They are terrified of becoming entrapped in their own ideas. It's not that Gemini aren't creative or brilliant; it's just that they're terrified of their own imagination's power. They're terrified of what they'll discover once they've found themselves.

They aren't frightened of their emotions, but they are always concerned about how they express them, about their words being misunderstood, and about unintentionally hurting someone's feelings. This is a common mistake made by Geminis as a result of how they externalize their fears. They are terrified of becoming engulfed in an emotion over which they have no control. Rather than feeling their feelings, they merely respond to them.

Their mind is a never-ending racetrack. They are continuously searching beneath the surface of their existing reality for something new. They aren't in a hurry to get someplace; they are simply looking for something fresh.

Perhaps the most crucial realization for a Gemini is
that there is no final destination at the end of the road.
They can't keep running forever. They'll have to take a
break and glance around at some time. To accept
responsibility for the environment in which they find
themselves. They'll have to examine the environment
they've created for themselves and decide if it's truly
what they want.

RELATIONSHIP COMPATIBILITY WITH GEMINI

Based only on their Sun signs, this is how Gemini interacts with others. These are the compatibility interpretations for all 12 potential Gemini combinations. This is a limited and insufficient method of determining compatibility.

However, Sun-sign compatibility remains the foundation for overall harmony in a relationship.

The general rule is that yin and yang do not get along. Yin complements yin, and yang complements yang. While yin and yang partnerships can be successful, they require more effort. Earth and water zodiac signs are both Yin. Yang is represented by the fire and air zodiac signs.

Aries and Gemini

A steady connection between Gemini and Aries has numerous potential. They can, however, thrive together with greater security if they both mature. They both place a high importance on freedom,

independence, and the satisfaction of gratifying their curiosity. The major difference: Aries' faithfulness against Gemini's infidelity. Overall, this is yang energy that is compatible.

Taurus and Gemini

When it comes to understanding each other's basic requirements, Gemini and Taurus have a challenging relationship. They have extremely different personalities. Gemini represents growth, movement, inconstancy, and infidelity. Taurus is the polar opposite. They will collide repeatedly, but if love is truly what unites them, they will be able to smooth things out and develop a solid relationship.

Gemini, and Gemini

A Gemini-Gemini partnership can work if both members focus on their complementary characteristics. They will have a lot of freedom. They will appreciate endless social gatherings where they can chat till they are weary. They will have a seductive and charismatic feeling. However, when problems emerge, no one wants to take care of their committments and responsibilities.

Cancer and Gemini

If Cancer and Gemini want to be together, they must be patient. Cancer is demanding because it is fearful. When asked for a lot of things, Gemini tends to flee. And Gemini isn't really interested in Cancer's excuses for its insecurity. A Gemini cannot be happy until they have independence. Cancer drowns in a glass of water if not protected.

Leo and Gemini

After smoothing out their inherent rough places, a Gemini and Leo partnership can be one of tremendous understanding. Gemini is a natural seducer. Leo, who is likewise a seducer, will not back its partner's efforts to conquer others. This may cause complications, as Gemini may feel suffocated by Leo's absolute exclusivity.

Virgo and Gemini

A partnership between Gemini and Virgo would be too hard to sustain and become steady because of the diverse ways they exhibit their natures. Both Gemini and Virgo are mutable signs ruled by Mercury, yet they work on separate wavelengths. Virgo is known as an

earth sign, while Gemini is an air sign. Gemini personifies dispersion and unpredictability. Virgo, on the other hand, does not leave anything to chance: it examines its steps indefinitely. This attribute is a hefty load for Gemini to bear on a daily basis.

Libra and Gemini

Gemini and Libra are two seducers that can get along without getting in each other's way. It is a connection that can blossom if they learn to compliment one another's strengths. Gemini has what Libra requires: the flow of energy required to move in the pursuit of greater pleasure. Libra has what Gemini needs: a more ordered existence with a better mix of feeling, thinking, and acting.

Scorpio and Gemini

If Gemini and Scorpio meet together just for the purpose of mutual intimate pleasure, it can be a beautiful experience for both of you. However, if you want to progress and create a relationship, you will have to walk a hard route together if you want it to succeed. Scorpio's jealousy and eruptions will destroy Gemini's chances of remaining by its side.

Sagittarius and Gemini

A partnership between Gemini and Sagittarius might be exciting if they are conscious that developing a lasting union requires certain sacrifices. Gemini and Sagittarius are both very gregarious, mobile, restless, and unstable signs. One of them should lead by example by becoming more solid. If they wish to avoid envy and friction, they must focus on their tendency to entice and dominate.

Capricorn and Gemini

Gemini and Capricorn are not meant to be together. Capricorns will spend their lives criticizing Gemini for its lack of concentration and for changing its mind so frequently. Capricorn's mental rigidity, on the other hand, will swiftly bore Gemini. Gemini will become impatient with the fact that it finds no way to really enjoy life with such a mate. It may work with hard work and a lot of love.

Aquarius and Gemini

Gemini and Aquarius make an excellent couple. They will understand each other almost completely. They are both gregarious and like spending time with

their pals. When Aquarius pays attention to others, Gemini may feel forgotten at times, but this will not jeopardize the connection. The emphasis will be on creativity.

Pisces and Gemini

Gemini and Pisces are more likely to split up than get married. Pisces will not feel comfortable in the company of someone as turbulent and unpredictable as Gemini. Pisces requires security and containment, which Gemini cannot supply even if it tried to. Pisces will be irritated by Gemini's lack of focus and propensity of running away from home. It is best to remain mute when it comes to integrity.

LOVE AND PASSION

Geminis are naturally gregarious and like meeting new people. They are romantics who like the start of new partnerships.

Of course, they know how to flaunt their attractions, as evidenced by the fact that many people regard them as wonderful company. They are very sexually active

and enjoy trying out new experiences. Sex is enjoyable for them.

If you fall in love with a Gemini, you become used to their mood swings. They can be aloof and unavailable at times, which perplexes their companions, who wonder why. Fortunately, these feelings are fleeting. The Gemini quickly returns their attention and displays their characteristic sense of humor.

Geminis like variety, which is why many of them struggle with monogamy. Some people are involved in two or more relationships at the same time. They normally mature and change over time, while others are unable to suppress their desire to be unfaithful.

MARRIAGE

Geminis are adaptable and cheerful by nature, and they make excellent partners, especially if their partner shares their desire for variety in all areas. Those seeking stability and routine, on the other hand, may find it difficult to keep up with them.

Geminis should think twice about marrying because it can take years for them to settle down and establish a commitment. When they meet the love of their lives, they should live with them for a while to see if they are truly compatible.

Those who marry a Gemini lady will be in the company of a creative and adaptable woman who wants to raise a family without sacrificing her job or her many interests.

When Gemini men decide to end their long years of bachelorhood, they make excellent spouses. Because they are thoughtful, happy, and smart, they are always happy and get along well with children.

CHAPTER TWO

GEMINI 2023 HOROSCOPE

Overview Gemini 2023

Jupiter will travel through your 11th house of Aries until May this year. After that, it will move into your 12th house of Taurus. This is not a very good transit because your opportunities will be limited. And in March 2023, Saturn will move from the 9th house of Aquarius to the 10th house of Pisces. This is a sign that Geminis will do well in their careers. Uranus spends the whole year in your 12th house of Taurus, and Neptune would be in your 10th house of Pisces.

Pluto would reside in your 9th house of Capricorn, however, in May - June of 2023, it will move to your 10th house of Aquarius. Gemini people would be

affected by these planetary movements all through this year.

Those born under the sign of Gemini would have a lot of luck with love and marriage this year. There would be a lot of anger and other strong feelings. Like never before, people would want to be with you. You win them over with your charm, especially your wit and sweet tongue. Mars and Venus, the love planets, are moving in your favor this year.

This year, big changes could happen when Saturn moves into your 10th house of career in March. You'd have a lot of energy, and you'd have a lot of work to do. But Saturn would make it hard for you to do well at work and hold up any professional goals you had for the year. Getting along with bosses, business partners, and coworkers may be problematic. There would be problems along the way, so not everything would go as planned. Face problems and stick to your position.

This year would be good for the health of Geminis. You would have good mental and physical health all year long. When Jupiter moves through the 11th house, there is no sickness. Saturn is also good for your health, but it will limit your energy. Be careful not to get diseases that are contagious. If you ate well and worked out, your health would be better for the year. Keep your immune system strong throughout this time. Because

Mars is in the wrong place, there may be times when people feel tired. Stay upbeat and be positive.

As 2023 begins, Geminis will be doing well financially. There would be a lot of money coming in. But you might also have to pay for things you don't want to, which could strain your finances. Geminis are told not to spend too much money during the year. Spend on what you need, but not on what you want. Jupiter's good effects on the fourth house would help you make money through land deals.

Geminis would have good travel prospects in 2023. As Saturn is in the 9th house of long-distance travel, natives can expect to take some trips abroad this year. If you live far away from your home country, the first half of the year is an excellent time to return. After the first quarter of the year, Geminis will be able to take many short trips that are both profitable and fun after the first quarter of the year. Be ready to go on unexpected trips as well. The cost might burn your fingers, so be prepared.

Geminis would do well in their spiritual pursuits in 2023. You would learn more about spiritual things as Saturn moves through your 9th house, and Jupiter gives you a year of faith, devotion, and piety. Worship your Creator every day and look for ways to make him happy. If you can, try to fast every once in a while. This will give you both mental and physical energy.

Geminis should reshape or remold their ideas if
they hit a dead end this year. You're about to have a
time of success, so use it to strengthen your personal
and professional connections. If you want to be
successful in your career, you need to take advantage
of all the opportunities that come your way. Be patient
and accept whatever the situation or environment
offers without getting angry.

January 2023

Horoscope

You evolve in a pleasant climate. Mars is in your sign, and on the 13th, he will start moving straight again. He helps your wishes come true because he has a good connection to Jupiter. As for the energies that come from Aquarius, they help make everything almost perfect by bringing just the right amount of change. Neptune in Pisces and Venus, who joins him at the end of the month, are the only shadows on the board. These energies want to give you false ideas, but you should stay in touch with reality so your brilliant ideas can come to life.

Venus in Aquarius frees you from feelings that stop you from being yourself when it comes to matters of the heart. Your loved ones feel better all of a sudden.

Love

Venus is in Aquarius from the 4th to the 27th, allowing you to be liberated from emotional shackles

that dampen your spirits. Avoiding the unknown and learning how to be romantic are two of its many benefits. You'll be amazed at what you can do in this oppressive atmosphere.

You're available and generous to those already in a relationship with you. You're pleased to fulfill the wishes of your better half.

Every light is lit up so that a new beginning might take place under ideal circumstances for singles. Make a little effort to keep it going at the end of the month if you want it to last. Be more romantic and talk about love more often.

Career

In the following weeks, you'll receive job offers from friends and acquaintances that align with your talents and experience. In addition, you will be able to surpass and exceed yourself as a result. There is no reason for you to lose out on these opportunities when the climate is conducive to your growth. Make sure you know what you're listening to if you want things to change.

You could see a bump in your professional possibilities if you're lucky this month. Get ready for a great future! You'll have a lot to cope with, but you're up to the task.

Your bosses will be aware of your hard work. However, you'll have to watch out for your coworkers, as this could be seen as a form of competition.

It's time for job seekers to step up their game. You'll be able to demonstrate your cerebral prowess thanks to the stars. Maintain your enthusiasm and drive.

Avoid being overly competitive or egotistic about your worth. You may suffer from anxiety and despair if you overestimate your abilities, leading to a lack of confidence. There would be a great deal of time spent traveling, but it would be for nothing.

Finance

On a financial level, the way is clear. You can sleep easily. At the end of the month, if you continue to manage your budget carefully, everything will be fine.

This month, your financial prospects are pretty good because the stars are in your favor. There are clear signs that investments will make huge profits, so it's best to take advantage of any investment opportunity after doing a lot of research. The climate is good for business people to start new projects. So, you should get going on such plans.

Health

This month, the stars are in a way that is good for your health. We need only sound a note of warning against excessive workload. This should be strictly avoided, and energies should be spread out smartly to keep all normal activity going but not put too much stress on the system.

This can be done easily by making a new schedule of activities. There are some good reasons to take care of your teeth and ensure all normal precautions are taken. Aside from this, it's a pretty good month for your health.

Travel

This month, your chances of making money from travel are low because the stars are not in your favor. This month, you would probably travel alone mostly by train and road, with a fair amount of air travel.

Also, a trip abroad cannot be ruled out. But, certainly, these trips would not accomplish even a fraction of the goals. A good amount of this travel would not be related to your business or job. The most helpful direction would be East.

Insights from the stars

The last restraint goes on the 13th, and the horizon is open. You can complete your projects. On the other hand, take off your rose-colored glasses. This will keep you from making strategic mistakes. This month, you need to learn how to connect with people personally and professionally.

February 2023

Horoscope

You keep going in the same direction as last month! You have a lot of bright and ambitious energies that push you to make your plans come true. Happy opportunities come, and you meet people whose ideas you find interesting. You like to try new things and are working on several projects.

But if you want all of this to last, you must be realistic. Some ideas can be done this month, while others can't. Also, don't let the ideas of others fool you around the 5th, or you might have trouble around the 20th. The full moon and the new moon make you more generous but make you more likely to be let down.

Love

Venus in Pisces makes your life hard until the 20th. You feel like your loved ones are getting away from you. Your ability to attract people drops down! Don't

despair! When Venus moves into Aries on the 21st, you find your way and the people who are right for you.

Those in relationships have a hard time until the 20th. Your partner wants you to be there more. People say that you don't give as much love as you get. From the 21st, offer a nice thing to do together as a distraction.

For single people, Venus makes life hard until the 20th if you don't get to the point. If it's too hard to be romantic, you should wait. From the 21st, she puts you in touch with people who like you.

Career

Still, everything looks good. You get lucky, and you do everything right. When the situation calls for it, you use it to your advantage. If you happen to meet very strict people, you can make the situation more relaxed by being yourself. Don't listen to your imagination if you want everything to work out for the best of all worlds. Why? Because it would make you more likely to make mistakes that would ruin everything.

During this month, you have a good chance of moving up in your career, but if you're not careful, you could easily fall back a few steps from where you are now.

Finance

On the money side, this doesn't look good, but it's best to get back to the basics. You won't have to deal with embarrassing financial situations that put you on the spot.

If you have a lot of energy and things are going well, you could do very well financially this month. You would have the courage to stand up for what you believe in and the drive to go after what you want and succeed. You would get a lot of help from a good combination of events.

This month, you can expect to get a lot of money quickly. An investment would be a good idea. There is also a chance that your relationships with your bosses will improve to the point where you can expect to gain a lot from this. This month's environment would be suitable for making investments and starting new businesses.

Health

A fortunate set of circumstances favors your excellent health this month, so you have nothing to be concerned about. Those susceptible to chronic colds and mucus discharge would be much alleviated. As long as treatment is handled seriously, those with piles

can look forward to a period of alleviation and possibly full recovery.

On the other hand, all of this comes with a warning about the need for proper dental hygiene in everyday life. On this point, any lapse could lead to severe consequences. You should expect to be healthy this month, which is excellent.

Travel

A month when you can expect to make a lot of money because the stars are in your favor. This month, you will have a lot of confidence in yourself and the guts to make decisions. You would put a lot of thought into a travel plan that would make you a lot of money.

You would probably travel by train, car, and air a fair amount when you were on your own. A trip abroad is also not out of the question. Your trips wouldn't always be for business. You would achieve your goal, whatever it is. The North is the best direction.

Insight from the stars

You draw individuals from various backgrounds because of your excellent company. Keep your attention on those you truly connect with if you don't want to disgrace yourself. You will grow spiritually if

you work hard to improve your life and find your divine life purpose.

March 2023

Horoscope

You continue to grow in a favorable environment, but Aquarius's radiant energy eventually fades away. For luck to continue to be on your side, you must make wise decisions. This month, your goal is not to change your mind when your projects are experiencing challenges. The best strategy is to work through the problems rather than avoid them.

How? by focusing on projects that have a direct impact. In addition, pay attention to the guidance you'll receive starting on the 8th. No matter how hard it may be, always choose knowledge over expediency. Accept criticism around the 7th since it will help you grow. A close friend saves the day on the 21st of this month!

Love

The energies that come from Aries help your relationships. They start them off with many activities, trips, and projects. It's a beautiful world! Starting on

the 17th, pay attention to how you feel and try to calm down. This will keep you from being alone when you don't want to be.

Your partner can count on you to keep things from getting boring in your relationship. You are happy, but only as long as you don't only talk about your plans and goals. This would cause disagreements.

For singles, love can come out of the blue and lead to a passionate and exciting relationship. Don't quit because of a few minor problems. Instead, work hard, and everything will be okay.

Career

The move of Saturn into Pisces shows that there will be some small obstacles in this sector. But if you accept the challenge from the planet of wisdom, you can lessen its effects. Gemini, you have some good fortune in a lot of ways. If you want it to last longer, you must take it more seriously. Pay attention to detail. After that, you'll be happy to see that you indeed are a genius.

The horoscope has nothing particularly good to say about your career. There is a good chance that you will have a major disagreement with your bosses or partners. This should not be allowed to happen, and you should try to stop it from happening.

You might also be filled with a feeling of insecurity that would affect almost everything you do at work. You could try to fix the problem by switching jobs quickly or changing how your business works. This would be an awful situation to be. Any change should only be made after being thought about carefully. There would also be a lot of travel, which would be very useful.

Finance

Everything is fine from a financial point of view. Don't let how you felt on the 26th fool you if you want to keep this little miracle going.

A good month from a financial point of view. You can look forward to making lots of money quickly. Quite a few of you would also benefit from making investments. There's also a good chance that an old friend will do you a favor that could easily help you out financially.

Also, this month you'll find a way to deal with your bosses that will make the relationship very good for you. This could be a significant gain. Last but not least, associating with some smart, spiritually-inclined, and gifted people would be good for you both materially and spiritually.

Health

This month, the stars are in a good mood to help your health, so you should be in great shape for most of this time. Any tendency to get sudden, severe illnesses like fevers and inflammation would be greatly reduced. They would most likely not bother you at all.

This would also apply to people with any kind of tooth trouble. In fact, any problem with your dentures should be taken seriously and has a good chance of getting fixed. This is a good time for your health, and those who are already in the best of health can expect to stay that way.

Travel

This is an excellent month to make a lot of money from travel, as the signs from the stars are very good on this score. Those who want to study or train abroad or in a faraway place have a great chance of success.

You would usually travel alone by rail, road, and the air a fair amount. A trip abroad also can't be ruled out. Only a part of your travels would be for business. No matter your reason, your trips will get you what you're looking for. The best direction is south.

Insight from the stars

Even though this is not your thing, show some emotion in what you do and say. It is not much, but it will bring you a lot. Foreign gains will appear in your life. Keep taking calculated risks; your life will move forward more than you expected.

April 2023

Horoscope

For your part, your work and personal relationships benefit from the Aries energy. When it comes to novelty and change, your mind is open to it. Attracting people to yourself is a natural part of who you are. As soon as someone sees you, they feel drawn to you. You're in excellent spirits in this frantic environment.

All of your firepower is being utilized! Because of this, you are always thinking of new ideas. But if you want these advantages to work for you, you'll need to be realistic. You also need to put some restrictions in place. By being a little more moderate in your expectations, you will be able to avoid disappointment.

Love

This sector is maintained by your ability to divert attention when problems arise. Your charm begins to work wonders on the 12th. Your company is attractive and in high demand. However, even though you are in

a whirlwind of seduction that satisfies you, you may
feel lonely.

For those in a relationship, your projects or
friendships may take precedence over reason. In turn,
your other half may feel lonely and let you know.
How? By isolating you in silence to draw your
attention.

This month, singles, you multiply the meetings!
However, if you want one of them to move towards
stability, you must take the initiative. How? Don't
forget to send out small messages from time to time!

Career

You crave variety because routine bores you.
Unfortunately, you must deal with Saturn, who
significantly slows your progress. On your worst days,
you may feel as if you are on the bench. Gemini, luck
continues to smile on you, so take advantage of it rather
than trying to avoid what doesn't suit you.

There are very few promising signs for career
advancement in the astrological alignment you face
this month. You would work very hard, but your goals
would elude you. There is also the possibility of serious
disagreements with your partners or superiors.

This should be avoided at all costs, as the
consequences can only be disastrous. Try to anticipate

difficult spots and work your way around them. There would also be significant travel, resulting in no gains, though a trip to the south might yield some progress for you.

Finance

On the financial side, there is a soft heart behind the rough exterior that wants to please. If you don't want to get in over your head with your finances, set a limit and don't go over it no matter what.

This month, your financial prospects look pretty good and could put you on a solid financial footing for the long term. You can look forward to getting a lot of money quickly. Some people would make money through stock trading, which would also make them a lot of money.

You would learn how to deal with your subordinates or employees in a way that lets you get the most out of their work. This would be a big plus for you and help you make a lot of money. And lastly, your relationships with your bosses would become so pleasant that you would benefit greatly from them.

Health

Lady Luck is in the mood to bless your health, so you can expect to stay healthy this month. People with

chronic illnesses like rheumatism and similar problems like flatulence and too much gas in the digestive tract can expect to feel better if they take care of themselves. This is also true for any kind of tooth pain.

You can also expect any nervous tendencies to improve and cause less trouble than usual. You might feel weak, but this is easy to fix with a bit of exercise and good food. A good month in which you probably won't face any serious health problems.

Travel

A month when it doesn't look like you'll get much out of traveling because the stars aren't in a good mood. You would probably travel by train, car, and air a fair amount when you were by yourself. Also, a trip abroad is not out of the question.

All these trips might be related to work and other things in equal measure. But no matter why you're doing these things, it's almost certain that you won't achieve even a fraction of your goals. As a result, it's a good idea to go over your travel plans ahead of time to see if they'll get you anywhere. The best direction to go would be South.

Insight from the stars

You get along well with other people and your group. But you sometimes have to take the lead if you want to keep good relationships. This month, you will make progress in your career. Everything is going in the right direction, so you should be happy with yourself.

May 2023

Horoscope

This month, Jupiter's time in Aries is over and you have until the 16th to fill your address book and your schedule with opportunities. Then you will be able to take advantage of all these benefits with the help of Sun in Gemini and Mars in Leo.

Even though you have a lot of energy and drive, take some time to think. Sort through everything you have collected to keep only what is realistic and doable.

Don't forget that Saturn's dissonances are keeping a close eye on you. They're not there to make you unhappy; in fact, that's not why they're there. They put roadblocks in your way to stop you from doing things that wouldn't be in your best interests.

Love

Until the 7th, you decide what happens. Your loved ones meet your expectations. Then the atmosphere could become less lively and lose its charm. If you

want to avoid these problems, try to find magic or romance at candlelit dinners as much as possible.

This month is a little chilly for Geminis who are in a relationship. Even though you have enough money to keep going, this might not be enough. To break the ice, tell your partner what you want and how you feel.

Singles, this month, some meetings are canceled, and others are set up. This process can make you feel uncertain or alone. Use this time to consider whether you want to be in a long-term relationship.

Career

It looks like May will be a good month. You experience ease every day. You are doing a great job reaching your goals. But, starting on the 17th, you can feel a sudden stop. Don't move on when this happens. Use this time to review everything you've done in the past few months. After that, you can focus on what's most important, and your plans won't fall apart.

A month isn't good for advancing your career and would also be a sign to be careful as there is a good chance that you will have serious disagreements with your partners or bosses. As much as possible, this should not happen. Try to be patient and stay away from situations that might cause trouble.

There would also be a lot of travel, which seems like a waste of time and money, though there is a small chance of getting something out of going to the west. It would also seem pointless if a lot of hard work didn't lead to any results. You might be tempted to break the law to make money quickly in such a situation. Stop doing these things firmly if you don't want to invite trouble.

Finance

Your romantic relationships can cost you a lot of money. So, don't be stingy with your money, but try to control how much you give and how happy you feel.

A beneficial month from a financial point of view. You would make a lot of money quickly, which is likely to happen. Others would make a lot of money from trading activities, and you would be able to see quick, valuable results from your efforts.

Health

This month, the stars on your right side have a lot of good news for your health. Any tendency to have one tooth trouble or another should become much less bothersome. Be careful not to overdo it, though, because that could easily ruin a good situation. Make a new schedule that doesn't stress your body much.

If you don't take care of this, it could be terrible for your health. The rest is fine. People who are already like this wouldn't be bothered by a tendency to get nervous. A pretty good month in which you probably won't face any serious health problems.

Travel

The horoscope from the stars says nothing particularly good about the benefits of travel. This month, almost half of your trips would be for work or business, and the other half would be for other reasons.

You would probably travel alone most of the time, primarily by car or train. A trip abroad is also not out of the question. No matter why or how you travel, you likely won't get even a fraction of what you planned to get out of it. Thinking carefully about your travel plans before you make them would be wise. The best way to go would be to the west.

Insight from the stars

Your favorite planet, Mercury, wants you to think about everything that has happened in the past few months. Do this even though it is boring. It will be worthwhile. This is an excellent month to invest because your money is set, and you can only pay for

what you need during the Mercury retrograde dates in 2023.

June 2023

Horoscope

This month of June is very private, but it's more boring regarding your job. Saturn in Pisces causes dissonances that make it hard to stay on track. Unfortunately, you're not very motivated because you're bored. Even though this isn't the best answer, change your mind as soon as possible. Venus and Mars in Leo work together with Mercury and the Sun in Gemini to make this little trick work. Your life is made more interesting by these sparkling energies. They like it when you talk with your group. They add movement to your life by taking you on vacations or other trips. Also, they tell you to make the most of the relationship that Jupiter gave you a few weeks ago.

Love

It's still a bit boring until the 12th, but don't worry, you'll be able to make up for lost time! Mars and Venus in Leo give your loves lively energy that makes you

feel good. Your family and friends make you want to go out, date, and hang out with friends.

Even if you're in a relationship, you can't do much until the 12th, no matter how good your intentions or care. You're bothered by something. Then things improve. You do a great job of putting people at ease and making the atmosphere happy and pleasant.

If you're single, you're feeling a little sad. You're not sure if you want to travel or meet new people. Lucky for you, your friends help you find your way out and learn how to seduce.

Career

Saturn has taken over the wheel of this sector, which is not a good thing. You find the time passing slowly because your promises have yet to be fulfilled. Do not believe you've been cursed! Instead, go through everything that has been presented to you. There is always an opportunity in all of this. You can make it come to life if you want.

This month's star alignment does not bode well for your career prospects. There is a good chance that you will have serious disagreements with your superiors. This would be a disastrous development. As a result, you should work hard to avoid such an occurrence.

Furthermore, you may be burdened by a futile sense of insecurity. This may prompt you to seek redress by changing jobs or business operations. Travel will also yield no benefits, though a trip north may yield some.

Finance

You have until the 11th to update your financial accounts. Do it! Even if you find it annoying, you will not be sorry.

Your financial prospects are excellent this month, as the stars appear to be on your side. You could be a stone's throw away from a sudden gain and not realize it. You will reap a bountiful harvest of unexpected gains this month and will also be able to reap immediate benefits from your efforts.

There is a chance that investments will also be profitable. Most importantly, circumstances will emerge that will allow you to treat your superiors in a way that will be highly beneficial to you. Association with some wise learned people would be equally helpful.

Health

This month, the stars are on your side and will shower you with good health. People who often get

cold hands and feet would notice a significant change, as their hands and feet would become noticeably less clammy.

Any persistent tooth trouble would also cause far less bother and, if treated diligently, would have a good chance of being cured.

There is also some solace that predisposition to nervousness and related disorders would be significantly alleviated. Overall, a favorable month in which you are unlikely to face any serious health risks.

Travel

This is not a good month for travel, according to astrology. Any pilgrimage to a holy place would be delayed or stopped by problems. Unfortunately, your dedication would not get you through.

Those who wanted to get a higher education or training abroad or somewhere far away would also have to deal with tough problems. You would travel alone mostly by car and train, with a fair amount of time spent flying. A trip overseas is also a possibility. But these trips would be utterly useless. The most favorable direction is Northwards.

Insight from the stars

It's time to get to know the new people you've met. One of them might be able to help your career move forward. Make your family happy by taking care of things. Always know that your friends and family will have your back.

July 2023

Horoscope

Your primary focus should be on achieving your goals. But the dissonances caused by Pisces and Virgo prevent it from succeeding. When you're feeling these energies, it's hard to know whether or not to have faith. They trick you into thinking you want greater independence, just to change your mind when you ask for it.

You don't want to feel like you're tethered to the ground! The temptation to make a dramatic decision is understandable, but it's not the best option. Waiting while using your relationships to attain your goal will save you a lot of headaches. The best way to deal with stress is to get out and meet new people. Reconnect with the people you've known for a long time. So, by doing so, your thoughts will become more lucid, and the answer will come to you at the perfect time.

Love

Your ability to stay within the boundaries Mars in Virgo places on you starting on the 11th will be critical to the success of your romantic relationship. The planets in Leo promise a month filled with intense feelings and heartfelt vows of love if you abide by their terms.

Bored? Use your imagination to keep yourself entertained. Having a relationship with you is a piece of cake. Don't allow yourself to become enamored with grandeur to the point that you begin to doubt your own happiness. Stay calm and collected, and everything will work out just fine.

When it comes to those looking for a partner, you will meet an attractive individual if you accept an invitation or go on a trip. Creating a connection is the only way for things to progress.

Career

This month, there is little to cheer about. You're more bored than ever by the same old thing. You get the impression that each day is the same. The moment you decide to make any headway; you'll find yourself right back where you started. Gemini! If you don't like it, don't let it stop you from achieving your goals. As long as you're at it, don't drift away.

This month's horoscope does not bode well for your professional aspirations. If you're feeling good about yourself, you might consider changing jobs or making other significant alterations to your business or service. Additionally, you may likely have disagreements with those in positions of authority.

This should be avoided since it will only worsen your current predicament. To make matters worse, you'd have to put in a lot of effort, leaving you frustrated and unsatisfied in the end.

Stability is preferred over erratic behavior, so strive to maintain it. People in your life may not be pleased to watch you grow. Be wary of their words of wisdom since they may want to bring you down or wish for your destruction.

Finance

The money is slipping through your fingers in the financial department. Unless you intend to become broke, set up a budget for each aspect of your life and try to stick to it.

You'll have a fantastic month ahead of you when the stars align, thanks to your positive outlook and boundless energy. As long as you have the confidence to stand up for what you believe in, you will be able to pursue your dreams and make them look effortless.

In this case, you would benefit much from a series of fortunate circumstances, and you can look forward to a bountiful crop of unexpected riches. Profits from investments would be substantial as well. Even more likely, you will be able to benefit much from the good connections you have built up with your superiors. This is a great moment to invest and start new businesses.

Health

This month, the gods are on your side regarding your health, and you will likely avoid any major health issues this month. You would no longer be susceptible to spells of sudden acute sickness, such as fever or inflammation, and these issues would no longer be a problem for you.

There are, however, reasons to be wary of your oral hygiene. Any lapse on your part here could lead to dental issues. Bone injuries, which are highly unlikely this month, should also be treated cautiously.

Travel

The stars aren't exactly aligned to make this a great month to travel, so you won't get much out of it. While traveling, you risk suffering an injury or experiencing

some other form of bodily discomfort. Because of this, you should exercise caution and take only the smallest possible risks.

You'd prefer to travel alone and primarily by rail or road, with some air travel thrown in for good measure. Traveling for work or business may account for a portion of your journeys. A sizable chunk of the population isn't very closely linked. It's safe to say that they would be a waste of time. The possibility of a journey abroad is not ruled out, and the most advantageous direction is East.

Insight from the stars

Stop thinking about what's bothering you for a while. Your daily life is greatly improved by the stars in Leo. Be ready because the unexpected can happen and change how things are going. You and your partner will grow closer and eventually get the hang of it and enjoy each other's company.

August 2023

Horoscope

Saturn's dissonances make your life harder, and Mars in Virgo makes this even worse. You are really stuck this month. Your goals are hard to reach right now, but that doesn't mean they're impossible.

Mars wants you to act in a methodical way if you want to reach your goals. Mercury is also in Virgo, so you must pay attention to the little things. Saturn tells you that you will reach your goal in time. Use the energies that come from Leo instead of trying to control fate or events. Get in touch with your coworkers, friends, and people you care about. Put your ideas out there and listen to what others have to say. When Mars moves from Aries to Libra on the 28th, things calm down.

Love

Your actions are nice, but it's not enough. Things are still having trouble getting better. This month, you

should go out and meet new people. Go see your friends. Accept invitations and go on trips over the weekend or on vacations. The blockages go away at the end of the month.

If you're in a relationship, you should stay reasonable even if you don't like how things are. Venus likes the way you feel about your partner. She wants you to spend time together in a place where you can talk about how much you love each other and have fun.

Venus promotes and encourages dating and flirting for single people. Where? by yourself or with friends. Wait until the end of the month for things to move forward. Show yourself off in the best way you can for now.

Career

Even though it's a good time to relax and do nothing, this sector worries you. It seems like nothing is going in the right direction. On the worst days, you feel like nothing is moving forward. Problems never happen by themselves, so some people help you figure out what to do. Sad to say, you don't agree with what they say. Don't be afraid. A friend will offer you something exciting.

The way the stars are lining up this month doesn't give you much hope about your career. You would

constantly worry about your job, which would drive you crazy. This would change everything you do for your business or job.

It's possible that you could work very hard to reach your goals, but they would still elude you. You would look for a new job or make significant changes to how your business works. All of this, though, wouldn't make you happy. There is also the chance of getting into a big fight with your bosses. As much as possible, you should try to avoid this. Try to think ahead and take care of problems before they get too bad.

Finance

When it comes to money, daily life is secure, but if you want to make a significant investment, you should take the time to research the business you are about to partake in.

This month, you have great opportunities to make more money because the stars are in your favor. Musicians, actors, painters, playwrights, and other artists will likely have a very good month financially and creatively.

In fact, things would be going very well, and you can expect to make a lot of money quickly. Investments would also be helpful and make a lot of money. There

is also a chance that a female friend will help you out, which could be a good thing for your finances.

Health

This month, your health will likely be in good shape because of good circumstances. People prone to long-term illnesses like rheumatism and digestive problems like flatulence and too much gas would feel much better. This means you wouldn't have to worry about these problems if you take standard precautions.

But if you have a sore throat that doesn't go away, you should be careful. This should be looked into carefully to see if there are any health issues, and then it should be treated with care. If you don't do this, it could mess up a good health situation that is going well. Aside from this, you don't have any real reason to worry.

Travel

This should be a good month for travel, and you should make good money from it. Most of the time, you would travel alone by train or car, with some flights thrown in. A trip overseas is also not out of the question.

Since not all of your trips would be related to your job or business, you would have a lot of success in getting what you set out to do. You would meet a lot of new people and have a lot of new opportunities. The best direction to go would be Northwards.

Insight from the stars

As of now, the stars in Leo are beneficial to your everyday life. As a result, making attempts to alter the situation is unwarranted. While you're rethinking your ideas, wait. When it comes to pursuing your aspirations, it's never too late. Put your faith in your abilities to make things happen.

September 2023

Horoscope

Pisces and Virgo clash, making it difficult to make progress. Several impediments stand in the way of your success. Discouragement can come out of nowhere. Do not be deceived by these dangers; they are not there to make your life unpleasant. It is best to let go of your hopes and dreams to avoid repeated disappointments. Wait for things to work out rather than trying to rush them.

With the help of Leo and Libra's energies, you'll be able to better connect with others. You should put your efforts here if you want things to get better. Why? Meetings and talks can lead to the development of a project.

Love

It is a good month for you and those you care about the most. You're a success in all you do. That smile of yours is magical. Your followers are desperate to get

your attention, and they'll go to any lengths to get it. Listening to the thoughts and feelings of others opens up a world of possibilities.

Even though Mars and Venus favor your relationship, Mercury in Virgo is a thorn in your side. The promise of the Moon this month will not alleviate your other half's resentment. Showing that you are reasonable can help relieve tensions.

Singles can keep meeting new individuals as long as they want! You flirt with a ferocity that reflects your personality. If you want a meeting to develop into anything more, you need to give more attention to the person you like rather than treating them the same as everyone else.

Career

Your progress depends on how often you can take stock and not let things get out of hand. Even though it's not in your nature, Gemini, you should think before you act. Before you send a file, make sure you haven't forgotten anything. Write down everything you need to do in a notebook. When it seems stuck, don't keep going because you think it will work out on its own. Find the answer.

According to the horoscope, there isn't much good news about your career this month. There is a good

chance that you will have major disagreements with your bosses. You can avoid this by trying to find trouble spots ahead of time and finding ways around them.

Also, this month would be filled with a feeling of insecurity. This would change the way you work as a whole. You would be unstable if you tried to change your job or how your business works. Any change you make should be done after careful thought. You would also work very hard, but even so, it's likely that you wouldn't reach the goals you set out for yourself.

Finance

On the financial side, it's the same fight, but with more details. Don't buy anything too expensive, because that will put you in the red.

According to your horoscope, this month should be financially good for you. Writers, poets, and other people like them would have a particularly good time financially and creatively.

You would literally make a lot of money quickly. Business and trading would also bring in a lot of money. Also, your relationships with your bosses and partners will likely improve to the point where you can expect to gain a lot from them. Also, the climate could

be very good for making investments and starting new businesses.

Health

This month, the stars in front of you are good for your health. People with sensitive chests or lungs prone to health problems in these areas are likely to feel a lot better. There is a chance that overworking yourself will make you tired and weak.

You could easily avoid this if you didn't push yourself too hard. Once this is done, everything will be fine. This would also help you deal with the slight chance that you have a nervous disorder, though it's not likely. If you take care of yourself, you can stay healthy for the whole month. Pay a little more attention to how well your teeth are doing.

Travel

The horoscope from the stars doesn't tell us anything particularly good about what you can gain from travel. Writers, poets, and other people like them might not have the best trips. In fact, some of them could be seriously hurt because their stays aren't productive.

You would travel alone mostly by train or car, but you would also fly a fair amount. A trip abroad is also not impossible. But it's very unlikely that even a holiday, which might not be all that fun, would come out of these efforts. The best way to go is to the East.

Insight from the stars

Don't run away at the first problem. Instead, you should stop and think about it. When that is done, fix what needs to be fixed. You will have to learn to be patient with yourself this month. Do not be in a hurry. Let things work out the way they should.

October 2023

Horoscope

You have the impression that your room for maneuver is shrinking by the day. As a result, it might make you resentful and drive you to make judgments based on the wrong motivations.

A negative attitude toward others or bothersome events will not get you anywhere. Instead, remain a safe distance from the situation. You'll be able to see things in a new light if you do this. As a result, you will act in a different manner. Mars, Mercury, and the Sun in Libra will assist you in this endeavor by fostering friendships and promoting new knowledge while highlighting your strengths. This month, put your lightness to one side and focus on strategy.

Love

Your charm never fails to work. You always get what you want. You can be harsh for the wrong reasons, which is a shame. If you want your love life to

look its best this month, show your friends or spouse how much you care. Instead of criticizing them, do what they want you to do.

Your relationship is affected by the needs of your family. Things won't get better if you run away, though. This month, the balance of your relationship will depend on how well you can put up with things that bother you.

For single people, it's easy to meet people, but if you want things to go further, you have to accept some limits. This month, you will meet someone with a future if you enjoy the charm of family life.

Career

This month is not a happy time. You think it takes too long. When things go wrong, you think you're just unlucky. Gemini, don't let fate hit you! For now, you need to use what you have and make it work. How? What? By thinking about what has happened in the past. Tell yourself that this time, which you believe to be so long, is actually your best friend, despite what you might think.

Your professional prospects are not looking too good this month because of the stars. There is a good chance that you will have serious disagreements with your bosses. You must not let this happen and work to

stop it from happening. Try to guess where problems might occur and go around them.

During the month, you might feel insecure, which could cloud your judgment. You could choose to switch jobs or business operations quickly without a good reason. Any change should only be made after careful thought. Traveling won't help you in any helpful way, but a trip north could bring you some small benefits.

Finance

On the money side, living from day to day works out pretty well. This month, think about your expenses more often and save money.

This month, your financial prospects look pretty good because the stars are in a good place. There is a chance that your female friend will do you a good turn, which could be a big help.

The month looks pretty good, and you could expect to make a lot of money quickly. Even trading and gambling would be profitable. Also, there are signs that your relationships with your bosses would improve, which would be very good for you. The environment would also be suitable for new businesses and investments.

Health

This month, the combination of stars in front of you is good for your health. If you have a stomach and digestive organs that are easily upset, this will help a lot. So will long-term chest problems like asthma, coughs, and colds.

You should be careful about the health of your teeth because this would bother you, but if you take good care of your teeth, nothing terrible will happen. Also, there are some reasons to think that you might be irritable and a little bit upset most of the time. Stay calm and balanced. You can keep your mind and body in excellent shape with a little exercise.

Travel

The stars don't look suitable for travel this month, so there aren't many chances of making money. This month, you would probably travel alone, mostly by car or train, with a few flights thrown in.

Also, there is a chance that you might work or travel abroad. But it is almost certain that these efforts would not bring the expected profits, pleasure, and satisfaction. The best direction would be to go North.

Insight from the stars

You like to live at your own pace and without problems. Sad to say, it tricks you. This month, you should be happy to be limited in ways that make you want to run away for the wrong reasons. No matter how small an accomplishment is, you should be proud of yourself.

November 2023

Horoscope

Even though you are stuck, a way out is slowly becoming available. How? By getting help from people you know that like you. Sadly, if you mess up, you won't get anything out of it.

This month is an excellent time to take advantage of your relationship without being sneaky. To do well in this little trick, don't give in to the urge to make everyone your friend. Instead, think things through. Find the people who can help you or who have interesting ideas. Make real relationships with them. To help you, think about Venus in Libra and how she deals with people and subtleties.

Love

Your relationships grow in a romantic environment. Your partner or your fans would do anything to make you happy. When Mercury moves into Sagittarius on

the 11th, things get more complicated. Changing what is being said won't solve the problem. On the other hand, it will be wiser to show that you are confident about the future.

For people in relationships, even though you try hard and have good intentions, conversations take a more spiteful turn. Anything can be a source of disagreement. To avoid this, you should occasionally feel romantic. Your significant other will be happy.

Singles, When Venus is in Libra, dating is easy as early as the 9th. Your charm works wonders. On the other hand, don't say what you think about relationships because it could turn off people looking for a serious relationship.

Career

If you only think about what's wrong, you will make the wrong choice. So, instead of going over all the mistakes in your work, look at all the good things about it. Have a look around. As you know how to do so well, make happy connections with other people. And if you have brilliant ideas, make them happen. By doing this, you'll get rid of this habit that makes you want to go as far as you can.

There's nothing particularly good about the signs for your career this month. There is a good chance that

you would have major disagreements with your bosses. No matter how hard it is, you should do everything you can to avoid this happening because if it does, it can only hurt your career.

Throughout the month, you would also have a deep-seated feeling of insecurity, which could affect how you act professionally. You might choose to switch jobs or business operations quickly. You should only make a change after carefully thinking about it.

Finance

On the money side, being balanced depends on the choices you make. So, weigh the pros and cons before you make a decision.

The stars say that this month's turn of events would not be suitable for your financial growth. Yes, there are clear signs that some of you would lose a lot of money if you invested. So, it would also be wise to avoid all kinds of gambling.

There are also signs that any dispute or lawsuit you might be involved in would almost certainly go against you, causing you to lose a lot of money. So, it would help if you tried to ensure that the decision in any such matter is put off until a better time. Relationships with business partners are also likely to go downhill. If you

can't stop this from happening, you could lose a lot of money.

Health

This month, the stars are in a way that is good for your health. People prone to sudden, severe illnesses like fevers or inflammations would feel a lot better, even if they only last a short time. Most likely, this kind of trouble wouldn't bother you at all. The same thing would be true for back pain.

But there are reasons to worry about the chance of an eye infection. This might bother you for a short time, but you could avoid even this by staying clean and taking a right preventive medicine. Overall, for your health, that's excellent news for this month.

Travel

Nothing particularly good about what the stars say about what you'll get out of traveling. This month, you would travel alone mostly by car and train, but you would also fly a fair amount.

There is also a chance that you will travel abroad, but these efforts won't bring the profits or pleasures hoped for. This is a pretty dark picture, but it's true. A

lot of your travel may not be necessary, and you could do just fine without it. The best direction to go is West.

Insight from the stars

Novelty solves your problems. This month, you will get out of a rut or get what you want by making real connections with other people. Be willing to make changes that will improve your life and the lives of those you care about.

December 2023

Horoscope

This month, the dissonances from Pisces are made worse by the ones from Sagittarius. They make you feel like you've come to a dead end. This feeling, which should make you unhappy, makes you want to get rid of it at any cost. To do this, you will think that any means are good! Some of them will actually be good. Others won't have the same luck, though. It's a good idea to find your freedom to act. But try not to use plans that will work against you. Instead, it would be best if you focused on opportunities that will lead somewhere. Things will take longer to set up if you take the serious route. But in exchange, you'll avoid avoidable disappointments.

Love

Venus in Libra keeps things calm and peaceful. With the dissonances of Sagittarius on the 5th, the

weather could become more unstable. This month, don't make your loves more exciting by making them compete with each other. Why? Because it will lead to arguments and never-ending requests.

For people dating, things can change quickly when Mars is in the picture. This might make you laugh at the time, but this little game could also lead to unpredictable reactions. Don't go too far this month, and everything will be fine.

For people who don't have a partner, Venus moves into Scorpio on the 5th. Even though it doesn't change much about you, it makes you harder to seduce. Your way of doing things could work, but only if you don't take advantage of it.

Career

Before making a decision or starting a project this month, you should consider it. Why? Because you can be led to false ideas that keep you stuck and make it hard to get out. Gemini, for once, don't believe everything you hear. And if you say you'll do something, make sure you do it. By doing this, you'll save yourself the trouble that would have saved you time instead.

This is a good month to get ahead in your career. People who like art and people who work in the fine

arts can look forward to a time when their creative work will be very satisfying. Some of you may even make a name for yourselves with your profession.

There would also be a lot of travel, which would be helpful. The best direction would be to go south. Besides traveling, you might also change where you work or do business. But think carefully before making any changes because a hasty move could easily undo a lot of your hard work.

Finance

Check how much money you have before you use your credit card!

This month, your financial goals should be easy to reach, according to what the stars tell you. You would literally make a lot of money quickly. Investments and trading business would also be a great way to make money.

Writers, poets, and other creative types can look forward to an inspiration that will be both financially and creatively helpful. Having friends who are wise and smart would be just as beneficial. Also, there is a good chance that your bosses, business partners, or employees will help you out in a big way. Your relationship with them would change for the better and grow into something good.

Health

This month, you have a number of good things going for you that are good for your health. Chronic diseases like rheumatism and gout, as well as problems with the digestive system like flatulence and too much wind, would get a lot better. This shouldn't be taken as a free pass to stop being careful, though. With regular care, there would be a relief.

There are some reasons to be a little worried about the state of your teeth. If you take care of your teeth, nothing terrible will happen to them. In fact, you have a good month ahead of you in which you won't have to worry about any significant health risks.

Travel

According to the horoscope, this is a good month to travel because the stars are in a good place. Travel can be good for writers, poets, and others like them, both in making money and getting ideas for their work.

You would usually travel by yourself, mostly by car or train, with some air travel. A trip abroad is not impossible. You can be sure that these efforts will help you reach the goals you set out to reach. The best direction is to go South.

Insight from the stars

Don't take the emergency exit if you feel like your freedom is being threatened. Instead, try something new and go through the main door! Single people will finally find their true love. After looking for a long time, you will find that one person who makes your heart full.

Printed in Great Britain
by Amazon

Printed in Great Britain
by Amazon

83458060R00169